Routledge Revivals

MODERN SCIENCE

MODERN SCIENCE

A GENERAL INTRODUCTION

BY

J. ARTHUR THOMSON

M.A., LL.D.

PROFESSOR OF NATURAL HISTORY IN THE UNIVERSITY OF ABERDEEN

Routledge
Taylor & Francis Group

First published in 1929 by Methuen & Co. Ltd.

This edition first published in 2018 by Routledge
2 Park Square, Milton Park, Abingdon, Oxon, OX14 4RN
and by Routledge
711 Third Avenue, New York, NY 10017

Routledge is an imprint of the Taylor & Francis Group, an informa business

Publisher's Note
The publisher has gone to great lengths to ensure the quality of this reprint but points out that some imperfections in the original copies may be apparent.

Disclaimer
The publisher has made every effort to trace copyright holders and welcomes correspondence from those they have been unable to contact.
A Library of Congress record exists under ISBN: 29030415

ISBN 13: 978-1-138-55523-5 (hbk)
ISBN 13: 978-1-138-56774-0 (pbk)
ISBN 13: 978-0-203-70556-8 (ebk)

MODERN SCIENCE

BY THE SAME AUTHOR

WHAT IS MAN?
SCIENCE AND RELIGION
TOWARDS HEALTH
THE WONDER OF LIFE
THE BIOLOGY OF THE SEASONS
THE SYSTEM OF ANIMATE NATURE
THE NEW NATURAL HISTORY

HUNGER AND LOVE, NUTRITION AND REPRODUCTION: THE MAIN MOTIVES OF LIFE

THE GOLDEN EAGLE BRINGING FOOD TO ITS YOUNG

MODERN SCIENCE

A GENERAL INTRODUCTION

BY

J. ARTHUR THOMSON

M.A., LL.D.

PROFESSOR OF NATURAL HISTORY IN THE UNIVERSITY OF ABERDEEN

WITH SIX PLATES AND TWENTY-NINE ILLUSTRATIONS IN THE TEXT

METHUEN & CO. LTD.

36 ESSEX STREET W.C.

LONDON

First Published in 1929

PRINTED IN GREAT BRITAIN

PREFACE

THE aim of this book is to give a general idea of the way in which modern science looks out on the world. By selecting a few salient illustrations it seeks to show how the various sciences are disclosing the Order of Nature. It is intended especially for senior classes in schools, and it is meant to be of particular use to scholars who are specialising in one branch, or who are not going forward to scientific training, yet would like to know what modern science in general is trying to do. But it is hoped that the book may also be of service to the able-minded reader who wishes an introduction of an informal type to the chief scientific problems of to-day. The book is meant to be suggestive as well as informative; and two characteristic features may be noted, for they are deliberate: the illustrations of scientific progress that have been selected are taken from all the great orders of facts,—from astronomy to anthropology; and they deal not with easy things, but with the big problems that matter most.

J. ARTHUR THOMSON

THE UNIVERSITY
ABERDEEN
July, 1929

CONTENTS

LIST OF PLATES

LIST OF ILLUSTRATIONS

MODERN SCIENCE

CHAPTER I

THE MAKING OF WORLDS

SCIENCE tries to reach back to beginnings, but it can never reach *the* beginning. For if we picture a scientific beginning, that is to say, something simple in the way of matter and energy and mind, it is always open to some one to ask : But what was there before *that ?* If the answer be there was nothing before that except itself, then we are saying that we can picture what is *everlasting*, which is far too daring a thing to say. If the answer be that before what we can picture there was a simpler something that we cannot picture, then we are confessing that we have not got back to the beginning. Thus science, as science, does not speak about *the* beginning, nor about *the* end either.

But it is a very different kind of question to ask : What was the beginning of the Solar System, with its sun and planets ? ; or How did the Earth begin ? ; or How did Flowering Plants begin ? ; or How did the Flying Birds begin ? These are reasonable scientific questions, to which we can give some sort of answer, which becomes more and more satisfactory as the years pass. For " knowledge grows from more to more."

An old naturalist once showed us a small island in a noble river, and told us with some pride that he had been an eye-witness of its whole history. It began from a couple of big trees that had been undermined and swept down by a flood. They stranded in a shallow and formed the beginning of a barrier that grew and gathered soil. After long years the outcome was a substantial island with trees and shrubs, even with birds and beasts of its own.

This is but an instance of what is always going on, for the face of the earth is continually changing. Great changes follow the widespread cutting down of timber in times of war ; the bed of a river sways from side to side ; the coast-line, even a very rocky one, is always altering ; villages that were once on the shore may now be miles inland, and, on the other hand, fields where the cattle once grazed are now good fishing grounds. We have to think also of gales, sand-storms, forest-fires, landslips, floods, and other violent influences that have often greatly altered the face of a countryside. Much more important, however, than sudden catastrophes are the imperceptibly slow changes of climate, such as those that brought about the Ice Ages, and the great changes of level, such as those that lifted up the Himalayas or made an island of Australia. How well Tennyson put it—he was always happy in his references to science :—

> There rolls the deep where grew the tree,
> O earth, what changes hast thou seen !
> There, where the long street roars, hath been
> The stillness of the central sea.
>
> The hills are shadows, and they flow
> From form to form, and nothing stands ;
> They melt like mist, the solid lands ;
> Like clouds they shape themselves and go.

It is very plain that the earth has not always been as it is, and what stories the geologists tell us of moulding and sculpturing, of washing away and depositing again, of breaking down and building up ! We must get accustomed to the idea that the outermost crust of the earth has been unmade and remade, over and over again.

But where the geologists leave off, the astronomers take up the tale, and lead our thoughts back to the young earth —to a rotating gaseous mass at a high temperature, in some way or other heaved off from the parent sun.

> They say,
> The solid earth whereon we tread
> In tracts of fluent heat began,
> And grew to seeming-random forms,
> The seeming prey of cyclic storms,
> Till at the last arose the man.

There can be no doubt that our earth arose from the sun, and a good case can be made out for the theory that it arose as one of the knots on the arms of a huge twisted nebula. The other knots on the arms became the other planets, and perhaps the earth knot was originally a double knot, including the moon from the very first. It is possible that the nebulous arms were drawn out from the gaseous sun by the tidal attraction of a passing star ; so that in a somewhat fanciful way we may think of the earth and the planets having two parents, the one being the sun, the other a great unknown. We should notice here that our present-day sun sometimes shoots out flaming solar " prominences " to heights of 100,000 miles, and some authorities would multiply this figure by three. These are often well seen when there is a total eclipse of the sun.

The theory is that on the solar arms or out-drawn swirling gaseous masses there arose at intervals the condensations or knots that we have spoken of, and that further ingathering of each of these knots formed the liquid or solid cores of the future planets, including the earth. As the condensations continued, each core gathered in by gravity what was left of the huge arms between the knots, besides immense quantities of fine dust-like particles which have received the long name " planetesimals." In some such way arose the separate planets, with our earth amongst them. Venus and Mercury are nearer to the sun ; Mars, Jupiter, Saturn, Uranus, and Neptune are farther away from the sun. Between Mars and Jupiter are the minor planets or asteroids, the largest probably not exceeding 300 miles in diameter, and these may be residues that were not concentrated into one body. No doubt these theories are beset by many uncertainties and difficulties, but there is general agreement among astronomers that our solar system arose from a vast nebulous mass, whose centre formed our present sun, while out-lying secondary nebulæ, which we have spoken of as " arms," were *somehow* drawn out, and then *somehow* condensed into separate planets, each revolving on its own orbit, and each with its own rate of rotation on its axis.

So we trace the earth back to its parent the sun, or more accurately, to the nebula whose condensed centre eventually became the sun we know. But this inevitably raises the next question : Where did the solar nebula come from ? The usual answer is to point to some of the nebulæ observable in the heavens to-day, especially to what are called the spiral nebulæ. Each is an enormous whirling mass of gas with a central nucleus and with spirally-twisted ejected arms emerging symmetrically from opposite ends (see Plate I). Condensations in the arms of these whirling nebulæ are stars being born. If these condensations should begin to move as detached bodies, clusters of stars might arise. Thus the astronomer traces back a great system of stars, like the Milky Way, to a gigantic rotating nebula. "In the spiral nebula," Sir J. H. Jeans writes, "we are watching not the birth of planets, but the birth of the stars themselves." And our own sun is but a small star.

The general conclusion is that our particular solar system had its beginning in a great nebula which formed many other suns besides. The subsequent formation of planets was by a different method, as we have already noticed. What is hinted at to-day in some of the spiral nebulæ is rather the making of stars or systems of stars, than the making of planets from a sun.

It seems that the universe of stars is a finite flattened Milky Way or galaxy in which the number of stars diminishes with increasing distance in all directions. It is a little like what we see when we walk out from a city on a very dark night, the lamps become fewer and more distant as we pass into the country where there are none, except perhaps at occasional houses. The total number of stars within the city and its suburbs—the whole galaxy in short—is estimated at about 47 thousand millions. Many of these stars are in systems of their own, some comparable to our solar system, while others are very different. These systems of stars show movements as wholes as well as movements among their members. That is to say, there are big stream movements through space, as well as revolutions and rotations.

Besides the main galaxy or Milky Way, there are on the distant outskirts what are called "globular clusters" of stars, and still farther into the dark country are the "non-galactic" spiral nebulæ which are believed to be new systems in the making. The question now rises: Can we get any farther back in our attempt to picture the making of worlds?

No doubt our sun was once vastly large, before it became the parent of its brood of planets; no doubt the moon was once born of the earth, or arose, according to another

PLATE I

DIAGRAMMATIC SUGGESTION OF A NEBULA WITH TWO SPIRALLY TWISTED ARMS AND TWO CONCENTRATIONS

theory, as a dwarfish twin-sister; perhaps a star passes through a kind of life-history, from brilliant giant to dark dwarf; and perhaps the numerous double stars may once have been single; but the fact is that the astronomers seem very unwilling to lead us back to any actual beginning of things. So far they are willing to go back—to the origin of the earth and the moon, or of solar and stellar systems, for hints of similar origins are to be seen in the heavens of to-day; but beyond that most of them confess that they have no clue. In fact they lead us back to a universe not very different from that which at present exists—as if the world was *an eternal now.* Thus Dr. Dingle writes in his brilliant book, "Modern Astrophysics": "Every avenue back into the past shows a universe sensibly identical with the universe of to-day. There is no record of a different world." As Browning said: "After the last returns the first, though a wide compass round be fetched."

The same confession of ignorance must be made when we look forward. "It is impossible," Dr. Dingle writes, "to forecast even the most probable destiny of the universe." No doubt there are stars and other bodies that seem to be passing through a cycle of changes. What one "heavenly body" is to-day,—to-day for the observer's eyes, perhaps many centuries ago for the observed body—may be very different when several ages have passed. "Unless some accidental catastrophe should occur, Betelgeuse will be, in the next stellar age, what Aldebaran is in this."

But if a giant star shrinks and degenerates into a cold dwarf, what becomes of it then? This seems an almost unanswerable question. We do not know what may happen. The dwindling stars themselves do not give us much information; they are too faint to be seen! They are *inferred*, as we might infer the presence of a man from his shadow. The number of dark bodies in space does not seem to be very great—as stellar numbers go. Perhaps it is unnecessary to suppose that stars actually die out. Perhaps they explode before they die—explode into dust and vapour, bringing our thoughts back to a diffuse nebula again. "After the last returns the first." Becoming, being, having been—and then becoming again. Scientifically, then, we cannot picture either the beginning or the ending.

We must not think that the science of the making of worlds, often called cosmology, consists of guesses at truth. It is based on what is observed to be going on just now, and astronomy is one of the most precise of the sciences. Wild theories do not last long. Just as we argue from what goes on to-day on the surface of the earth to what may have gone on millions of years ago, so we trace the earth back to a parent sun, and the sun back to a nebula which formed clusters of stars. For we see nebulæ making star clusters to-day. But can we not get back farther still, behind the nebula ?

If the cosmologist is pressed very hard he will perhaps go a little farther back to a time when the matter of the universe was evenly distributed through space, something like what exists to-day in the " cosmic cloud " of atoms in inter-stellar areas. The original state may have been a universe of widely separated or " highly dissociated " atoms, which slowly gathered into nebulæ. Sir Oliver Lodge has calculated that if all the matter in the universe were uniformly spread out—it being supposed that the universe is finite and measurable—the molecules would be about five yards apart, whereas there are a million million million molecules in each cubic inch of the air we breathe. It has been suggested that this uniform distribution of molecules of matter became disturbed, and that millions of vast regions of greater concentration arose thousands of millions of miles apart. These were the primitive nebulæ which formed the clusters of stars !

A little child becomes a wise man, gradually and naturally, and no one doubts that there was in the beginning the promise and potency of the outcome. So whether we start with a universe of highly dissociated atoms or with a nebula, we must hold firmly to an idea stated long ago by the clear-thinking Aristotle, that there is nothing in the end which was not also *in kind* in the beginning. We well know that there is mind in what is for us at present the end, namely Man, the minister and interpreter of Nature. Must we not therefore come to an old conclusion, which we cannot more than dimly understand : In the beginning was Mind : all things were made by it and in it was life, and the life was the light of men.

CHAPTER II

THE IMMENSITY OF THE UNIVERSE

IN our first study we asked the very difficult question, how our solar system may have come to be as it is, and how the system of stars to which our sun belongs may have come into being. Let us now think of the worlds as they are, and of their immensity.

Our forefathers were thrilled by the apparently boundless and unfathomable sea and by the apparently unending plains, and we are starving ourselves to-day if we are not gaining many such impressions of the "wide, wide world." Who can forget the first climbing of the hills and the first sight of another county, stretching into dimness, and beyond that the sea joining the sky. But modern man has annihilated terrestrial distance; he can journey round the world, even without the help of wings, in about sixty days; he can send a whisper round the globe in fewer minutes.

So we have to turn to astronomy—the science of the "heavenly bodies"—to get something of the old thrilling impression of immensity; and in astronomy the difficulty is that the distances are often so great that they fail to move us at all. It makes little practical difference to us whether the astronomer says millions or billions. In America and in France a billion means a thousand millions (10^9); in Britain it usually means a million millions (10^{12}); but what difference does it make to us, who cannot picture what even one million means. It is useful, then, to think for a little of some of the devices by which the immensities of the world may be made more picturable.

When the astronomer tells us that the earth is about 93 million miles from the sun, around which it revolves, whereas the moon, the nearest of the heavenly bodies, is only about 240,000 miles from us, we cannot but appreciate in some measure the difference between the two figures, but perhaps we gain more if we say that the light from the sun takes eight minutes to reach us, whereas the sunlight reflected from the moon reaches us in less than a second and a half.

Light travels at the rate of about 186,300 miles a second, and nothing travels any faster. Thus it is very convenient to speak of a " light-year," the distance that light travels in a year—namely 6 million million miles. The nearest star is Alpha proxima, and it is about four " light-years " away. The next nearest, Alpha Centauri, is just a little farther away. Therefore, when we look at either of these stars, we are seeing the light that left it four years ago !

The light we receive from Vega started on its journey through space twenty-seven years ago. Most of the stars we can see with the unaided eye, we see by light that left them when Galileo Galilei studied them in the early seventeenth century, and with his new telescope discovered Jupiter's moons, and showed that the Milky Way is made up of clusters of separate stars. There are clusters of stars which we observe through a telescope by light that left them long before the Christian Era began, and there are spiral nebulæ in the distant heavens which are 10 million light-years away, say 60 million million million miles. Later on, other books must be consulted if one is to understand, as understand one should, *how* astronomers measure these vast distances. But that is beyond us in this book, which gives no more than glimpses.

An interesting consequence follows from the immensity of the universe, that it is only the near at hand that we can exactly study *as it is*. If we attend our minds to the earth at a given tick of the clock, then, as Dr. Dingle writes, " we must wait 8½ minutes before we can say how the sun appeared at the same moment, because the light of the sun takes that time to reach our eyes." Thus, by " the universe as it is," we mean " the universe as it appears." This difference is of little importance in connection with the general shape and position of the heavenly bodies, for these change very slowly as compared with the velocity of light ; but it has to be kept in mind when we are comparing near and distant heavenly bodies as regards their physical state, even their brilliance, for we are comparing them as they were at different times. When it is 1928 on the near one, so to speak, it may be centuries ago on the distant one !

But what we are thinking of just now is simply that the world-picture is on an immense canvas. According to the investigations of Dr. Hubble there are very distant nebulæ which are more than a hundred million light-years away. As we saw in our first study, they perhaps represent regions where the once much scattered matter of the universe became gathered together, and in the course of time became concentrated into glowing stars or suns.

Another yard-stick which the astronomers use is the "parsec," which requires a little explanation. This is given by Sir Richard Gregory in his extraordinarily lucid introduction to astronomy, called "The Vault of Heaven." As you stand in a field and look towards a tree a short distance away, it appears to be in line with others farther off; but if you walk a little it appears to be in line with different trees. "Just as a tree appears to change its place with respect to more distant trees as the observer moves round his path, so a star must appear to vary its position with reference to others deeper in space, in consequence of the earth's orbital motion." The star seems to move in a minute ellipse in the course of the earth's annual revolution round the sun. The angular measurement of half the longest diameter of this apparent ellipse is called the star's parallax. Thus a star might have a parallax of one second, one-360th of a degree, there being 360 degrees to a circle. Now if a star has a parallax of one second, it is said to be at a distance of one "parsec," using the first syllables of the two words, parallax and second. A parsec is about 19 million million miles, about three times a "light-year." The greatest diameter of the main galaxy that lies symmetrically about the plane of the Milky Way is said to be about 100,000 parsecs.

When Galileo in 1609 improved the telescope and turned it on the sky, he made many remarkable discoveries. We have mentioned two, that he saw four of the moons or satellites of Jupiter, and that he showed the Milky Way to be composed of a multitude of separate stars. An interesting change of outlook followed his discoveries, the universe began to loom much larger in men's minds. It had been regarded as bounded, but it began to appear boundless. As discovery succeeded discovery, the idea grew strong in the minds of astronomers that the universe was infinite.

Of recent years the pendulum has swung in the opposite direction, largely under the influence of Einstein's Theory. Science has gone back to the old conviction that the universe cannot be infinite, and attempts have been made to calculate its size. Its circumference has been estimated in a rough sort of way, by the Dutch astronomer de Sitter, at about a hundred millon light-years, or 600 million million million miles. This is ten times the distance from the earth to the farthest spiral nebula. The weight of this Einstein world of four dimensions is estimated at 10^{54} grams, or about a hundred trillion times the mass of the sun. One must try to keep in mind that in enormous stretches of the wide wide world, the matter is spread out very thin, perhaps about one atom to every cubic inch in the interstellar deserts through which starlight travels to our eyes. There are several thousand million million million atoms in a drop of water ; compared with that how thinly peopled is the space between the stars.

Another stupendous fact to be kept in mind is that most of the condensed parts of the universe are in a state as different from the earth as it is possible to conceive. For we are assured that nine-tenths of the matter in the universe is at an unthinkable temperature above 1,000,000° Centigrade, and must be in an extraordinary state of agitation. It must be a turmoil of maimed atoms and free electrons and mighty winds of ether-waves. This difficult description will become a little easier when we come to study the structure of matter.

Still more difficult, however, is to try to keep in mind the probability that the real universe is very different from our picture of it, and is as a whole quite unpicturable. For it is probably a world in four dimensions, somewhat warped or distorted by the presence of material bodies, as the surface of an india-rubber balloon would be distorted by a weight resting on it. We are bound to fall into error if we insist on trying to form an image of this four-dimensional world, for it cannot be pictured ; yet the supposition, in the hands of mathematicians and physicists, has already cleared up several dark corners.

Some people persist in fancying themselves journeying to the farthest nebula and then looking over " the edge of the world," so to speak,—looking over into a " beyond." If there is a circumference, they say, there must be something outside. " For leagues and leagues beyond, and still more sea," as Rosetti said. But the fallacy is in the assumption that we can possibly visualise the real state of affairs. We cannot.

Einstein's theories may be ignored or rejected, but if they are provisionally accepted, there is no sense in making a *fictitious* difficulty, as we do when we say that if we reached the farthest star of the most distant nebula we should still see " something " farther away, and so on *ad infinitum*. This is a difficulty that we make for ourselves by forgetting that we cannot visualise a four-dimensional curved world.

Suppose that a flattish caterpillar is moving in two dimensions, like a leaf-miner, in the skin of an orange. It burrows its way industriously to right and left, forwards and backwards, but it cannot eat its way *out*. As a matter of fact, many of the minute leaf-mining caterpillars do keep mainly to one plane. Suppositions are cheap when a start is once made, so let us suppose that the burrowing caterpillar had not only thought out some geometry, but was metaphysically inclined. Then it might say to itself : " I have been going straight ahead day after day, and I have never found the least hint of an end. Does this not mean that I am living in an infinite orange ? "

The caterpillar's " space " would be the rind of the orange —to its experience two-dimensional ; and the creature would be speaking quite accurately if it said that it never found an end to its world ; no, not after weeks and weeks. And it is important to notice that the caterpillar's straight lines would all be arcs of a circle, straight as they may have seemed to the caterpillar.

When the caterpillar duly became in the course of time a butterfly, and had experience of *three* dimensions, it would doubtless detect its previous fallacy and say : " What I should have inferred was that the orange was *unbounded*. It was, after all, a very *finite* orange, for I can now realise that I repeatedly made a tour of what seemed to me to have no end." And so for ourselves, if we could get free from the trammels of our three-dimensional experience we should perhaps find no difficulty in thinking of a universe boundless yet finite.

Perhaps all that can be safely said is that the idea of a finite measurable universe fits in better with what is at present known than does the idea of an infinite universe. Thus the experts say that the assumption of *unlimited* quantities of enormously distant matter originally distributed uniformly in space is inconsistent with the world as we know it now. And if the matter in the universe was once evenly distributed throughout all space, it is difficult to think of this space as other than finite. For if it were otherwise, would the molecules ever have come together at all in concentrated masses? Moreover, as a matter of fact the *observations* of the astronomers do not lend support to the idea that there are nebulæ beyond nebulæ indefinitely.

But it is high time to come back to certainties. The earth is 93 million miles or so from the sun, and it takes a year to complete its circle. Yet this great orbit might be included within the star Betelgeuse! There are visible bodies in the sky which the astronomer sees by light that left them 100 million light-years ago. We need not think that the immensities are the biggest facts in the world, for there is wisdom in the remark of the sailor-man, who, after seeing many of the wonders of the world, confessed that nothing had ever puzzled him so much nor had excited his admiration to the same degree as the finger-nails of a new-born baby.

One often sees very interesting news under the title "Island Universes." But this does not sound a very happy phrase. For if *the* universe includes *all*, as it is in duty bound to do, how can it be used in the plural? And how can a "universe" be an island, as if something not in the universe embraces it like a sea. The explanation or condonation of the phrase seems to be that the universe of to-day is so inconceivably vaster than that of two generations ago that it has been felt necessary to put it in the plural, and the word "island" is merely used to mean a well-defined system or aggregation of stars, more or less of a unity by itself, for instance, in having a revolution of its own. In short, "island universes" are very remote spiral nebulæ.

One of their features, so to speak, is their distance from us, for, if we use a light-year as yard-stick, the nearest fixed star is about four "light-years" away, and Sirius about ten, while, as one of the authorities tells us, "in order to reach the island universes we must go to millions of light-years."

Another feature is their immense size, for they may be tens of thousands of light-years in diameter, and may include millions of stars, all those that are individually visible being thousands of times brighter than our sun. "The best known of all island universes is undoubtedly the Andromeda nebula, a great spiral, just 1 million light-years away, and about 50,000 light-years in diameter. It contains millions and possibly billions of stars."

Another feature of some of the nebulæ is that they frequently produce new stars, that of Andromeda giving origin on an average to over two every year. The cause of these "novæ," as they are called, remains obscure, though the "novæ" themselves may be 10 million times brighter than our sun. The Magellanic Clouds which are much nearer us than the Andromeda nebula, being only about 100,000 light-years away, have all the essential features of a universe, but they are not known to show any appearance of new stars.

In a recent article on "Island Universes," by Mr. Luyten of the Harvard Observatory, another feature of these distant universes is noted—their speed in the line of sight. "The Andromeda nebula is approaching us with a speed of 200 miles a second, the Magellanic Clouds are receding from us at the rate of 170 miles a second. One spiral is even known to hurry away from us at the speed of 1100 miles a second. With reference to the whole system of spiral nebulæ, our Galaxy, and the sun with it, seems to be moving toward the constellation Cassiopeia, with a speed of 250 miles a second."

These are very wonderful facts, but even more wonderful is their discovery by man. Speaking for ourselves, we are convinced that half-way men or "tentative men," who afterwards become true men, emerged from a stock common to them and to the higher anthropoid apes, now represented by gorilla, chimpanzee, and orang, but when we reflect on man's *achievements*—for instance on his understanding of the astounding universe, we cannot think in any easy-going way of the manner in which he emerged. This impresses us here because we have just been trying to show how man has been a measurer of the universe. It is plain science that we must consider man in the light of evolution ; but we must never forget to consider evolution in the light of man. It was no "fortuitous concourse of atoms" that gave rise in the course of ages to the astronomers! Let us go back again to the Aristotelian saying that there is nothing in the end that was not also present *in kind* in the beginning, and say to ourselves under the star-strewn sky: If there is so much power of Reason in the astronomer's discoveries, there must also have been some kind of Reason in the beginning." "In the beginning was the Word."

CHAPTER III

THE SUN

IN our last study we thought about the largeness of the world in which we live. Our earth is a planet in the solar system, which is part of a larger system of stars; and on the outskirts of this vast system there are very distant nebulæ. Man lives in a little corner of an immensity, and yet man is its measurer. Man's mind has disclosed the orderliness, we might almost say the reasonableness, of the universe.

But we must now try to get a little nearer the heavenly bodies, and we naturally begin with the Sun, the centre around which the earth revolves.

There is a pleasant story of a London lady who asked the Parsee visitor if he was not one of those old-fashioned people who worshipped the Sun. "Yes, Madam," he replied, "and so would you, if you had ever seen it." That is, no doubt, the proper spirit, for though the detached astronomer tells us that our Sun is no star of stars, but "utterly mediocre and undistinguished," we know that the Sun is really greatest of all luminaries because it most directly lives again in us, who are probably the only creatures who hold the heavens in understanding. Not only does the Sun rule the solar system; it sustains all plant and animal life, and our own life, from day to day and from age to age. We cannot think of any embodied living creatures that are not built up of protoplasm or living matter, and that is only possible on a planet where water is present in a liquid state. For protoplasm does not contain less than 70 per cent. of water. It seems very improbable that there is any other planet on which there could exist living creatures such as we know; and there is no use speaking about others which we cannot even imagine.

The fact is that the Earth is a very exceptional place, very chilly when compared with a star, yet near enough the sun to be genially warm. It is the comparative coolness of the Earth that has allowed atoms to group together into little orderly clusters which we call molecules, and has allowed molecules to cohere into particles and form what are called the platelets and droplets of what is known as the colloidal state, of which we may take liquid gelatine, cod-liver oil emulsion, or living matter as familiar examples. Coolness and complication go together.

But the Sun is not only the sustainer of life, it is in a true sense the parent of the system of which it is the centre. As we said in our first study, the earth and the planets probably arose as knots on immense spirally twisted arms, perhaps drawn out from the Sun by the tidal attraction of a passing star. What is certain is that the Sun we see is the condensed centre of a nebulous mass from which the eight planets had their origin. So let us say it again, for it is a big thought: The earth and the planets must be traced back to a once larger Sun that was their parent, and to the children of one of its children the Sun gives food and clothing, health and wealth, and the light of the eyes.

To the very detached astronomer, as we have said, our Sun is but a phase in the history of one star out of millions of stars; yet even the astronomer, as discoverer, has to join in the universal obeisance. For the Sun is so near, as astronomers count distance—only 93 millions of miles away—that the investigator can peer into it, and submit it, piece by piece, to examination. "And not only so, but turning to account the very mediocrity of the Sun, we claim thereby to discover the constitution of millions of similar stars. The Sun is the great betrayer: it yields, not only its own secrets, but those of its neighbours as well." (Dingle's "Modern Astrophysics," 1924.)

As regards volume, or the space occupied, it would take over a million Earths to make one Sun, the diameters being 7918 and 865,000 miles respectively. But just as we are about to exclaim, "How enormous, then, our Sun," the stupendous astronomical multiplication table goes on. It would take 10 million times the volume of the Sun to make one Betelgeuse, which is the brighter of the two shoulder-stars in the constellation of Orion.

But though the Sun is about 1,300,000 times bulkier

than the Earth, it is only about 332,000 times heavier. That is to say, its average density is something like a quarter that of the Earth. The solar density increases from the surface to the centre, and as the centre is probably gaseous, it follows that the outer envelope must be very tenuous. Speaking of density and volume, we may notice here that the mass of Betelgeuse is only about 35 times that of the Sun, though its volume is 50 million times. It follows that the mean density of Betelgeuse must be much less than that of the Sun; it is not much more, indeed, than one-thousandth of the density of air. It is a remote star, one of the most massive of stars, yet in mean density one-thousand times more tenuous than the thin air, and of unthinkably enormous size! For Betelgeuse has a diameter of about 300 million miles. We may repeat from our second study the astounding fact that Betelgeuse is so large that within its circumference the earth could *revolve in its orbit.*

Compared with a star like Betelgeuse, the Sun is a "typical dwarf," that is to say there are many stars—sun and star being the same—with a diameter of about a million miles. Dwarf to the star Betelgeuse, giant to the planet Earth, the Sun is known to rotate on its axis, as is shown by the disappearance and reappearance of certain particularly well-marked sun-spots. The mean period of the rotation is twenty-five days, some hours.

The Sun's distance from us is known to be about 93 million miles; but this unthinkable figure leaves one cold. Can it not be brought nearer our understanding. Perhaps, as we said before, the best way is to remember that while light travels at the velocity of 186,300 miles per second, the sunlight we are at this minute enjoying left the sun eight minutes ago. But in Sir Richard Gregory's "Vault of Heaven" (2nd edition, 1924), perhaps the best introduction to Astronomy, there are two comparisons which help us to think of the Sun's distance from the Earth. A tour round the world, about 24,000 miles, can be made in some sixty days if all goes well. "To travel as many miles as separate us from the Sun it would be necessary to make nearly four thousand such journeys, and if a traveller started on his circuits as soon as he was born, he would require to live about 650 years to complete his task." A message travels along a nerve at the rate, for man, of about 400 feet in a second. Imagine a man able to stretch out his hand to touch the Sun ; it would take about forty years before he became aware that his finger was burning !

Around the Sun there is an atmosphere eight thousands of miles in thickness, so that the earth would be lost to sight in this outer envelope alone. Just as our atmosphere consists of oxygen, nitrogen, carbon dioxide, water-vapour, and rarer things, so the Sun's atmosphere includes many elements, and these tend to occur in zones or layers, though not in any rigid way. For terrific storms are often raging. Farthest out there is a calcium zone ; beneath that there are layers of hydrogen and barium ; beneath that there are other elements. Of the ninety-two terrestrial elements (four still undiscovered !) about forty have been demonstrated, by means of the spectroscope, in the Sun. Perhaps there is no element in the child (the Earth) which is not also in the parent (the Sun).

What then is the Sun's atmosphere ? A jostling crowd of atoms travelling at high velocities and often colliding. The collisions result in jerking off electrons. These, as we shall see in a later study, are unit charges of negative electricity, which, along with positive charges or protons, constitute the atoms of matter. Because of the collisions the Sun's atmosphere includes many free electrons and many *maimed* atoms, called ions. Surging through the jostling crowd there are great winds of ether-waves seeking for freedom. Some of them find a way out, and after a long journey through space they warm us and gladden us when we walk in the sunshine.

The Sun itself is probably gaseous throughout, the interior being under great pressure and in a state of terrific heat. The Sun's temperature is somewhere about 6000° Centigrade near the surface, but it soon rises to 1,000,000°, and reaches 40,000,000° at the centre. This unthinkable temperature implies an extraordinary velocity of "particles" ; the greater the heat the quicker the movement of the ions or atoms or molecules. The molecules in the air of a room are moving at the rate of about 500 yards in a second, but the atoms in the heart of the Sun must have a velocity of over 100 miles a second.

Sun-spots, with a dark centre and a lighter margin, appear suddenly on the Sun's face. They are often in groups and they seem to move across the disc, an appearance due to the Sun's rotation. They last for a variable time, from a few days to several months, and they seem to be due to internal eruptions of the Sun, producing a sudden uprush of gases. These expand in the rarer atmosphere, and the fall in temperature causes an apparent local darkening of the disc. They must be terrific whirlpools of atoms and electrons, and they sometimes have a diameter of 40,000 miles.

Very external compared with sun-spots are the "prominences," usually like stupendous red flames. They are short-lived or long-lived gaseous uprushes from the Sun's atmosphere, sometimes rising, it is said, to a height of 120,000 miles. They are flames of calcium, hydrogen, and some other elements.

A remarkable fact is that prominences and sun-spots attain a maximum every eleven years, which shows that they must have something in common, and that there is some orderliness even in the Sun's disturbances.

Above the Sun's ordinary atmosphere, there is a deep corona or chromosphere (5000 miles thick), best seen when the disc of the Sun is hidden by the Moon during an eclipse. "When the last remaining crescent of the photosphere is just obscured by the black disc of the Moon, the corona suddenly bursts into sight—a beautiful, pearly white halo." Part of the charm of this rarely seen corona or chromosphere is its mystery, but it seems to consist very largely of atoms of calcium (with one electron missing) which are able, as Professor Eddington puts it, "to float on the sunbeams." The sunlight travelling outwards is partly absorbed by the atoms which tend to fall inwards into the Sun, and those, like calcium, which are able to absorb large quantities of light in proportion to their weight, will be able to float most successfully. "The atoms in the chromosphere are kept floating above the Sun like tiny shuttlecocks, dropping a little and then ascending again from the impulse of the light."

The Sun is a gaseous world, becoming gradually smaller and denser. The great lantern of our world is a dwindling, darkening star. But when we try to picture the internal tumult of gases at a very high temperature, we must no longer think, as even astronomers used to think, of fuel burning away in a vast fiery furnace. If it were a case of burning away, the Sun's fire would have been out long ago.

Nor is it enough to suppose, as used to be supposed, that the Sun's heat is due to its continual shrinking. How, then, are we to account for the fact that the Sun has continued for hundreds of millions of years giving off annually an amount of heat equivalent to 120 billion tons of fuel. Where does all this heat come from ? In Professor Eddington's remarkable book "Stars and Atoms" it is explained how the source of the Sun's energy is to be found in the transmutation of elements, and the collisions of electrons and protons. Year after year in the laboratory the disintegrating radium-atom gives off supplies of heat, and some similar re-grouping of electrons and protons is probably going on in the interior of the Sun. It may be, for instance, that helium is being built up out of hydrogen, and if so, this would release surplus energy, providing an abundant supply of heat. Even more daring is the suggestion that energy may be supplied by the running together of electrons and protons in headlong collision. Mutually destroying one another they may form a splash in the ether which spreads in space as an electro-magnetic wave, meaning rays of heat and light to us. "By annihilating a single drop of water we should be supplied with 200 horse-power for a year." Fortunately or unfortunately man cannot as yet do this, but it may be that the Sun and other stars have the secret. In any case it seems almost certain that the chief fountain of the Sun's energy is to be found *within* its atoms or in the collisions of parts of its atoms.

The Sun is rotating on its axis ; around the Sun the planets revolve in their orbits. Nearer to the Sun than the Earth is, the planets are Mercury and Venus ; in order from the Earth outwards, the planets are Mars, Jupiter, Saturn, Uranus, and Neptune. Between Mars and Jupiter come numerous small asteroids. Each of the planets, like their fellow, our Earth, has its rotation and its revolution ; and most of them have one satellite, like our moon, or more than one. This makes up the Solar System, now known to be one out of a multitude. But besides rotation and revolution, we have to think of the mass movement of the Solar System as a whole, at the rate of many miles a second, towards an unknown goal in space, which is called "the apex of the Sun's way." We live in a little corner of the universe, but we are citizens of no mean city. Our ancestors in the days before Prometheus, before there was fire or any knowledge of the seasons, " lived like silly ants beneath the ground, in hollow caves unsunned," but we are becoming more and more able to enjoy and use and understand the Sun.

CHAPTER IV

STARLIGHT

IN our last study we tried to look at the sun, which is very blinding. It is easier to look at the twinkling stars, which our sun "puts out" by day. Yet, as we said, it is mainly through the sun that astronomers have come to understand the stars, for the sun is so much nearer that they can peer into it ; and then they can argue from the near sun to the distant stars. For stars are suns, and our sun is a dwarf star.

When we look up into the star-strewn sky on an unclouded dark night, we get an impression of countless numbers. But Spenser was right when he said in his "Færie Queene" that there were far more different kinds of animals in the sea than there are *visible* stars in the sky. "For much more eathe to tell the starres on hy, albe they endlesse seem in estimation, Than to recount the sea's posterity . . ." There are many tens of thousands of different species of animals in the sea—each itself and no other—but there are only about 5000 stars that are visible to the naked eye, and only about 3000 at any one time. It is difficult to believe that the number of visible stars is not greater than the number of people at a concert in a large city hall. But there is no doubt that our first unsophisticated impression that we can see many thousands of stars requires to be corrected, as can be plainly proved by counting !

Yet our first impression of numberless stars is in a way right, for with a field-glass we can detect in the course of time over 120,000 stars, while with a comparatively small telescope we can theoretically make out 7 millions. With a really great telescope, like that at Mount Wilson (with an object glass of 100 inches) hundreds of millions of stars can be seen. To which have to be added more millions which cannot be observed as such, though they are revealed by stellar photography. And so the number rises to about a thousand millions !

This is indeed a prodigious multitude of worlds, but modern research still further increases our estimate. We are clearly aware that the earth and the other planets form a system revolving around that parent star which we call the sun ; and that the sun is a member of a system of stars —those that we see clearly on a starry night. But, as we have already said, the modern view inclines to the conclusion that there are *other* systems of stars. For there are many nebulæ which are very far away compared with the stars of *our* galactic system, and each of these nebulæ is a vast system of hundreds of millions of stars. These distant nebulæ are very numerous (perhaps 2 millions of them) ; they are all somewhat similar in size ; and they are to a considerable degree evenly distributed through space. Thus in an astounding picture we begin dimly to realise that the earth is a member of a solar system whose central star is one of many in a vast stellar system, which is in turn one of many other stellar systems !

One of the finest of epitaphs is on the tomb of Fraunhofer, who helped to lay the foundations of spectroscopy. The simple words are : "Approximavit sidera"—He brought the stars near. The allusion is of course to the use of the spectroscope in analysing starlight, and thereby discovering the presence of certain metals and other elements in these distant worlds. For the light given out by different glowing or incandescent bodies varies according to the chemical nature of the body. Thus it is possible to bring the stars near. Some time or other the student should learn exactly how the spectroscope works, but that is beyond the scope of this book.

One of the investigator-geniuses of to-day, who has followed Fraunhofer in bringing the stars near, is Professor A. S. Eddington of Cambridge, and he has given us luminous pictures of what goes on in a star. Before the beginning of the twentieth century, no one knew the source of the light and heat that the stars give forth. They were supposed to be gigantic furnaces, stupendous crucibles, burning themselves away, and sometimes, as was observed, going out, as if they had exhausted all their fuel. One of the many difficulties in face of this theory, however, is that the stars have lasted so long. The true explanation, both for star and sun, came with Becquerel's discovery of radio-activity, for, as was noted in our last study, the fountain and origin of the greater part of the radiant energy of the stars and the sun is to be found in the re-arrangements and perhaps collisions of the electrons and protons that build up all atoms. The power of the light is in a deep sense the liberated energy of the dust!

From the nature of the light received from the stars, the expert is able to calculate the temperature of these glowing worlds. So exact has this calculation become that he can tell us the difference between the temperature of the surface and that of the centre. The surface may have a temperature of 3000° Centigrade, while the internal temperature may vary from 2 million degrees in outer zones to 20 millions at the centre. This means that the particles are rushing about at a terrific speed and frequently colliding with one another. Perhaps we should say "relatively terrific," for while the speed of the atoms of helium at a temperature of 4 million degrees Centigrade must be about a hundred miles a second, the laboratory physicists work with helium atoms emanating from radio-active substances at the rate—sustained for a very short time—of 100,000 miles a second. So Professor Eddington actually speaks of "the jog-trot atoms of the stars."

In the inside of a star, then, there is a tumult of atoms rushing about in all directions, at relatively terrific speeds, and colliding with one another endlessly. The energy of their movements may be called "material heat," like that of a red-hot poker where the vibrations of the iron molecules are much more rapid than when the poker is cool. In certain conditions the poker will melt, a change of state more readily seen in the heated lead in the plumber's ladle, and the melting means not merely that the molecules of the metal are vibrating much more excitedly than when the metal is cold, it means that they are sliding over one another in actual locomotion.

But besides this "material heat" we must think of "ethereal heat," that is to say of electro-magnetic radiations, or, as they are often called "ether-waves." Like water welling out from a spring, so there is from within the star a surging of ether-waves, trying, as it were, to win their way out. But they are "encaged by the material," and it is only now and then that any one of them succeeds in dodging all the atoms and threading the maze into freedom. If they get free, we know them as starlight! We say "light," for in the course of its countless collisions with atoms, a ray which in the interior of the star begins with a short wave-length, gives rise to longer waves of visible light.

Gravitation is a name for the way in which pieces of matter hug one another; it is a gathering-together or condensing force, tending to form larger and larger masses. Thus the earth is always gaining meteorites, which we see as "falling stars" when it is dark. This ingathering or condensing force in a star, which is always drawing the atoms in towards the centre, is counteracted by the disruptive or out-driving force of the ether-waves, doubtless helped in some measure by the centrifugal force always developed in a rotating body. For the stars are rotating like our sun, which, let us repeat, we should call a star if it were not ours. Let us think then of this ethereal pressure working outwards, almost like a wind. It distends the star, working against the condensing gravitation, and it helps to bear up the weight of the outer layers.

When we ask where these surging ether-waves come from, we ask one of the most difficult of questions. Everything at present points to the general answer that they are started by re-arrangements of the electrons and protons that build up all atoms. As we mentioned in speaking of the sun, some experts believe that in the stars there may be a release of energy by some transmutation of elements, such as occurs in radio-active substances. As the atom of radium, for instance, disintegrates, there is a pouring forth of heat and light. The kind of transmutation that might perhaps occur in a star is the changing of hydrogen into helium, and this would liberate much energy. Another possibility that Professor Eddington speaks of is the generation of energy if an electron and a proton annihilated one another in a headlong collision. This is a very difficult subject, so we have repeated ourselves a little. What seems certain is that the stars are not to be thought of as mere furnaces. The probability is that the main fountain of starlight is from within the atoms, or from collisions of the electrons and protons jerked off from atoms.

We must turn for a little to a subject that will be clearer

after the study on the Structure of Matter. Besides the colliding *atoms*, about which there is no manner of doubt, and besides the surging *ether-waves*, there are in the star vast numbers of free *electrons*, that is to say unattached charges of negative electricity. Where do *they* come from? They have broken loose from the atoms, for when an atom tries to absorb or swallow a unit bundle or "quantum" of ether-wave, it is apt to burst; and that gives one of its satellite electrons a chance to shoot away on its own. But by and by a burst atom or maimed atom meets a loose electron, and "induces it to stay and heal the breach. The atom is now repaired and ready for another mouthful as soon as it gets a chance." So it goes on endlessly.

But we should notice that the atoms in the fierce world of a star are *always* being burst or maimed, so that the correct name for them is *ion*, which means an atom that has lost an electron or more than one.

What then are the partners in the hurly-burly of a star? There are (1) the maimed atoms or ions, (2) the free electrons, and (3) the ether-waves, which belong to what is called the X-ray group in our terrestrial experiments. The ether-waves are always, as it were, trying to escape from the star, but they encounter the obstruction of the atoms that absorb some of their energy. If the ether-wave makes its way gradually through the "atom cage," gaining in the course of time the boundary, it will dart outside and begin its long journey, changing into starlight! And this starlight is what we wished to understand even a little.

Stars differ greatly in density, for some are much more tenuous than the air we breathe, while others are so close-packed that a cubic inch of their material may weigh a ton. We read that "the companion of Sirius," which was for a long time unseen, "is composed of material so dense that a ton would be a little nugget that you could put into a match-box!"

Another fact to be thought about is that stars have what may be called a life-history. They seem to grow gradually into diffuse giants, with temperature rising, and then they sink into dense dark bodies, radiating into space only the energy that they receive from without.

When the part of the earth where London or New York stands has its face, so to speak, to the sun, the heavens are full of light unless it happens to be a very foggy or cloudy day.

When the rotation of the spinning earth carries us round twelve hours afterwards to a diametrically opposite position, it is pitch dark. We are supposing that the observer is not spending the summer in the Far North, where it may be daylight for three months or so, and we are also supposing that it is not a date when the sky is illumined by the sun-light reflected from the moon. This is very elementary, yet is it not worth asking the question why space should be dark at all if there are all these millions and millions of stars, most of which are pouring forth light ?

The main part of the answer is that the stars are so far away in the immensity of space that the amount of their light-energy that the earth intercepts is almost unthinkably minute, only one six-millionth part of our sunlight. Moreover, the light of a star cannot be reflected from other glowing stars in the way that sunlight is reflected from our dead satellite, the moon. The inconceivable distances of the stars should be recalled for a moment. For while the sun is 93 million miles away, and we see it by the light that left it eight minutes ago, we see the nearest "fixed" star (Proxima Centauri) by the light that left it about four years before. Most of the stars that we see with the unaided eye we see by the light that left them in the seventeenth century, in Galileo's day.

We cannot leave the stars just now without remembering that along with the march of the seasons, they gave our forefathers of long ago their first great object-lesson in the Uniformity of Nature,—which was in a way the *Beginning of Science*. As the great mathematician Poincaré has said in explaining what exact astronomy has meant to the intellectual advance of man : "Was I wrong in saying that it is astronomy that has made us a mind capable of comprehending Nature ; that under heavens always overcast and starless, the earth itself would have been for us eternally unintelligible." Man, with all his faults and failings, has this mark of nobility that he has been humble enough to *learn*. From the lore which the prehistoric Chaldean shepherds began to gather as to the movements of the heavenly bodies, there has grown the Astronomy of to-day, one of the intellectual achievements of which we should be proudest. We must continue "to hitch our wagon to a star," as Emerson said, for starlight has proved itself to be earth-light. We understand not only the earth but ourselves better in the light of the stars !

CHAPTER V

THE CHANGING STAGE

OUR studies began with an attempt to get back to the beginnings of worlds, and then we tried to take a bird's-eye view of the universe! Next came a study of our sun, and of other greater suns which we call stars. Now, naturally enough, we settle down to think about the earth. It would be a great gain if we could close our eyes and see—even in a general way—the appearance of the earth's surface age after age. This is one of the big views that we must try to get,—the changing stage on which the drama of life has been played during hundreds of millions of years. In a country like Britain we know that what we are familiar with to-day has not been for very long as it is now. Not very long ago there was no North Sea, and Great Britain was not an island. Not very long ago most of the country except the South of England was covered with thick glaciers, and this meant that all the larger animals were wiped out. There had to be a re-peopling from the Continent of Europe after the Ice Ages passed. It will take long study of Geology, one of the most fascinating of all kinds of study, before we are able to picture the appearance of the earth during the successive Geological Ages, which are represented by different kinds of rocks and fossils in the many-layered crust, but let us light the lamp of the imagination and see this big fact clearly: that the surface of the earth has been the scene of endless change.

It is not difficult to picture the first stage before living creatures appeared upon the earth. There was no eye to see, but if there had been a spectator, what would have been the spectacle? A monotonous smoking desert, cindery under foot, with no more scenery than sand-dunes. Here and there out of a crack comes a crawl of molten rock, like very coarse-grained tar, hardening and blistering on its surface as it cools, and creeping out in front from beneath its crust in an ugly sort of way. No sun by day, nor moon by night, nor any stars, but a thick curtain of cloud over everything and hiding the heavens. And beneath the cloud a dense and dusty unbreathable air, with carbonic acid gas, and water-vapour, and much nitrogen, but only a trace of oxygen, though it forms about a fifth of our atmosphere to-day. Thank the green plants for that! There was no sign of life at all, nor any sound save crackling

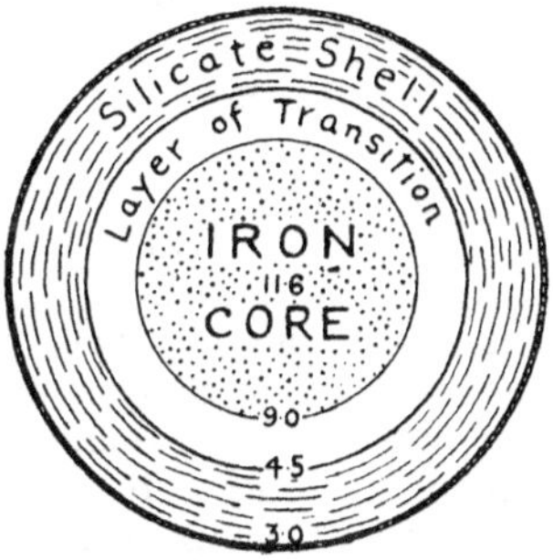

FIG. 1.

Section through the earth (by permission from Daly's "Living Earth"). The outermost dark line represents the crust, but is necessarily too thick in proportion. The figures 3·0, 4·5, 9·0, 11·6, give the approximate densities. The silicate shell is about 1000 miles thick; the transition layer of mixed iron and silicate is about 300 miles thick; the iron core is about one-sixth of the whole planet; the centre is 3980 miles from the surface.

and hissing, and now and then a big explosion. Such was the surface of the cooling earth—before it became the home of life—perhaps a thousand million years ago.

Ages passed and the crust had further cooled and hardened, sinking here and rising there, like a ball of clay in the potter's hands. In the depressions might have been seen, had there been eye to see, the gleam of water, gathered together from the precipitation of vapour on the cooled surface. Depths and shallows there were, the beginnings of seas and continents, but those authorities may be right who picture, at a very early dateless time, a universal ocean.

To any observer without a microscope the world of these ancient days would have seemed quite lifeless. Yet the seas must have been teeming with invisible creatures, hesitating between plants and animals, and, if we may judge from their successors to-day, many of them must have been very numerous, more in a cubic foot of water than we can see of stars on a clear night. Some of the most ancient rocks show traces still of these first living creatures, which are often called Protists.

Let us try to picture them,—invisibly minute spheres and ovals, propelled by living lashes, able to trap the sunlight that struggled through the clouds, and feeding on the almost fresh sea-water and on the air that it had entangled. In short, they lived on air, water, and salts, and on the energy of sunlight, just like our green plants to-day. If we could have been observers, we think we would have seen that many of these first organisms (living creatures) grew and multiplied by day, and died in the chill of their first or second night. For they had very slender resources. Life to begin with was like a flickering flame that spreads along the short dry grass, never rising high, never very strong, yet persistently refusing to be put out. Such were the first living creatures, perhaps eight hundred million years ago.

Ages passed, and by bucklings and warpings of the floor of the sea, the contrast was emphasised between the depths and the shallows, between the seas and the continents. In the inshore waters, shallow enough to be well-illumined, some of the primeval forms of life began to anchor themselves, becoming the first sedentary plants. Some grew into long threads, others into encrusting discs, others into branching fronds, forming the first fixed vegetation,—the primitive seaweeds. Taking due precautions, we should make a little party some day—we should never go alone—to explore the seashore rocks at the very lowest tide. There, on the wet rocks, before the flow begins, we should peer into the seaweed tangle, for this gives us a glimpse of what the earliest vegetation must have been, before there were any plants on the dry land.

Amidst the crowded ancestral seaweeds an epoch-making step was taken, for there appeared a new kind of life—the first animals. Whereas the seaweeds feed on air, water, and salts, the secret of animal life is to utilise the materials that plants have already worked up. Plants feed at a low chemical level—let us repeat "air, water, and salts"; animals at a high chemical level. Plants accumulate their riches from pence, animals start with the plants' pounds. Plants manufacture munitions, which animals steal and explode. This is one reason why animals are so much more energetic than plants, that their food begins where plants leave off. In other words, animals are predatory organisms, and we must think back to their beginning among the primeval seaweeds, where they learned to get outside of fragments which the waves broke off from the swaying fronds. Long afterwards they learned to eat plants whole, and to devour other animals. But in the last case, when animal eats animal, however long the chain may be, one must always come back to vegetation. In a deep sense all flesh *is* grass, and all fish is seaweed,—including, of course, floating as well as moored seaweed.

If we continued trying to get a glimpse of each successive stage in the history of the earth's surface, we should crowd everything else out of our book. So we must be content with two more pictures.

In the time of the laying down of the great coal-beds, which are now for the most part buried deep below the surface, there were immense forests in many parts of the world. The climate was moist and genial, and the vegetation was luxuriant. There were dense, dark, damp forests, swampy underfoot, and often penetrated by lagoons. But the forests were very different from those of to-day, since the trees were flowerless and seedless. The Carboniferous forests, as they are called, consisted almost wholly of giant horse-tails, tree-ferns, and club-mosses, propagating by means of spores, not by seeds. True trees they were, however, sometimes a hundred feet in height, and growing close together like the graceful horse-tails that we see to-day by the side of the marsh-pools. Often there must have been showers of spores sinking to the damp ground, like the so-called "sulphur showers" of yellow pollen that are borne by the wind from present-day pine-forests. On the swampy grounds there were somewhat newt-like amphibians, which spent the early chapters of their life in the pools. Many of them were armoured with scales, whereas the Amphibians of to-day are almost always scaleless. Some large, some small, they were the first backboned animals to get a footing on the dry land, if dry it could be called. They were the first animals to have fingers and toes, breathing nostrils and lungs, a movable tongue and a voice. Except, perhaps, for some insects, that made instrumental music cricket-wise, by rasping one hard part of the body against another, the silence of Nature was first broken by these somewhat unprepossessing pioneer amphibians that frequented the swampy Carboniferous forests.

Long ages passed, and in the Jurassic Period the stage of the drama of life had a very different appearance. It was in great part a time of widespread drought, with extensive tracts of dry steppe-land. There was a rich representation of reptiles, some of them giants, some of them dwarfs, very diverse in character. They tried all sorts of ways of living—crawling and running, climbing and burrowing, swimming and flying, just like their successors, the mammals, with which we are more familiar. On vast tracts of arid ground there must often have been a welter of reptiles—lizards, snakes, and tortoises, besides others so long since extinct, that we have no common name for them. But among these crowded reptiles we see with the eye of the imagination a very different kind of creature. It is a long-legged biped, about the height of a cock, very spare and lightly built. It has a long neck and a long tail, long legs and long toes. It is like a bird in its sharply marked-off head, its large eyes, its supple neck, its springy legs; it is like a lizard in its scales, its teeth, and its tail. As it sprints along the dry ground it flaps its fore-limbs, which bear a slight web and a covering of what look like scales partly shredded up! After it gets some way on, it takes a long running leap, skimming over the ground. Occasionally, when a big reptile makes a clumsy lunge at it, the high-strung startled creature leaps with a sharp cry on to a low-growing tree and disappears among the branches. This was the first bird—perhaps fifty million years ago.

As living always means a give-and-take between the creature and its circumstances—the organism and its environment—we see the importance of the changeful stage. The actors change the scenery, but the scenery also influences the actors. There could not have been life at all if there had not been liquid water, and there could not have been higher animals unless there had been dry land. The surface of the earth has been always changing, partly just because it is a mobile crust, which crumples and crumbles

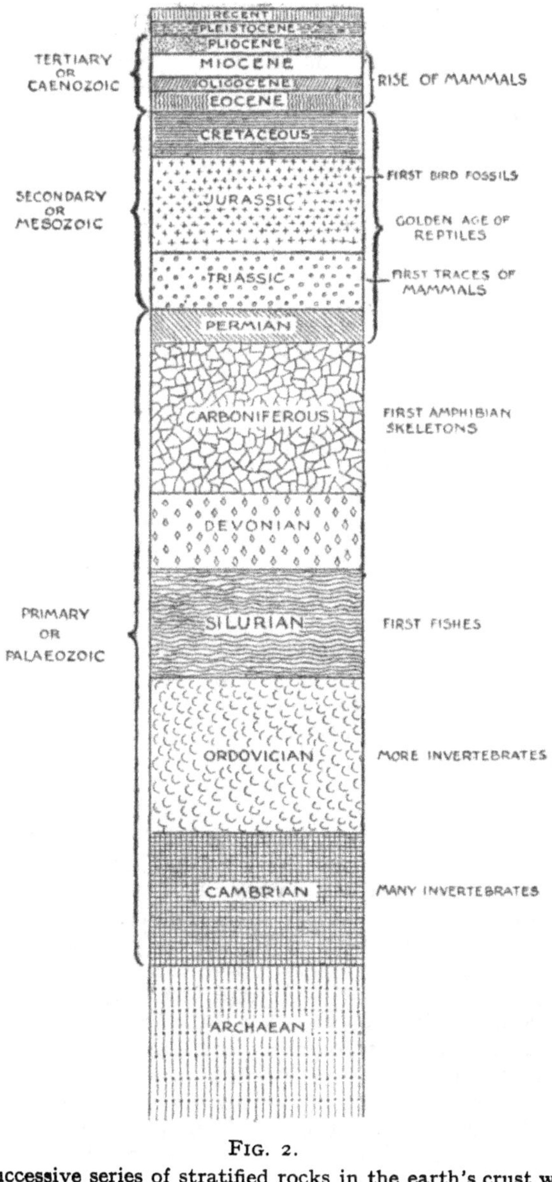

FIG. 2.

The successive series of stratified rocks in the earth's crust with an indication of some first appearances.

unceasingly, but partly because of the hand of life which is sometimes protective and sometimes destructive.

Let us think of the long-drawn-out drama of life, in the course of which finer and more masterful creatures emerged age after age, yet ever so gradually, for it may have taken a million years to fashion a feather. But let us think also of the great changes in the surface of the earth from age to age. The materials of the original crust, largely of the nature of granites, were partly crumbled into fine fragments by the slow but sure weathering of rain and snow, frost and wind, and by harder hammers like the stone-rolling river and the breakers battering on the cliffs by the seashore. What was broken into crumbs was laid down again as sediments, and pressed and heated into hardness, so that sandstones and mudstones and other layered (or sedimentary) rocks were formed. Animals built up shells of lime, using salts subtly filched from the sea, which owed them in turn to what the rivers brought down from the land. Vast deposits of these lime-shelled animals, small Foraminifera in particular, accumulated on the bed of the ocean, and were by and by lifted up to form chalk-cliffs, like those of Dover. In somewhat similar ways harder limestones were formed, and thus an endless un-making and re-making continued for unthinkably long ages. The same material was often used over and over again, and there were bucklings and dislocations and thrusts of the layered deposits, besides volcanic eruptions and outflowings of molten rock from the warmer zones under the cool crust. "O Earth, what changes hast thou seen!"

At no one place is there a display of more than a few of the great strata which form the sedimentary rocks—sandstones, mudstones, and limestones—of the earth's crust; but if they could be all seen together they would make a thickness of about sixty miles! It is plain that an inconceivably long time must have been required for their making. If a great cinema-film could be made, representing the changing surface of the earth, and the changing fauna and flora, and if proportionate lengths were given to the successive Geological Ages, and if the whole were arranged so that it could be unrolled in one day, beginning at ten o'clock in the morning, man would appear on the film a few minutes before midnight! And yet what the film would show in these few minutes would be in some ways grander than all that had been unrolled throughout the day.

CHAPTER VI

THE RADIANT ENERGIES

FROM our study of the heavenly bodies, to use the old phrase, we have come down to our own earth, where we are most at home. Here we must take stock for a little of the things and powers that exist apart from living creatures. And we cannot begin better than with the radiant energies which come to the earth from the sun or are resident in the earth itself.

It is often said nowadays that there are only three kinds of measurable things—Mind is not a *thing*—in the world, namely (1) *electrons*, each a unit charge of negative electricity; (2) *protons*, each a hydrogen nucleus, or, what seems to come to the same thing, a unit charge of positive electricity; and (3) *radiations*, which are usually regarded as electro-magnetic vibrations or as ether-waves of diverse wave-lengths. They travel through space at a uniform rate of 186,300 miles per second, the velocity of visible light, than which there is no greater velocity. The radiations produce various effects on the bodies that absorb them, thus some of them (called infra-red rays), that the earth catches, bring us warmth; and others that enter our eye give us light. But even this visible light consists of rays of different wave-lengths, which produce in us different colour-sensations. They can be separated from one another in a prism, and are so separated, as every one sees, in a rainbow. Those that produce the sensation of red have the longest wave-length, while those that produce the sensation of violet have the shortest. But just as there are sounds which we cannot hear—only a few people detect an ordinary bat's high-pitched voice—so there is light that we cannot see. Such is the ultra-violet light, whose rays serve as a tonic to the health both of man and beast. We are the better for them when we enjoy the sunshine. Unfortunately they cannot come through ordinary window-glass, but fortunately they can be artificially produced inside a building. An interesting point is that ants and bees can see them, though they are *invisible* to man. Invisible yet not inoperative, as is familiarly proved by their chemical action on the photographic plate, as well as by their tonic effect on the living body.

We have already stated the interesting fact that many of these radiations are started in star and sun by convulsions inside atoms or by head-on collisions between electrons and protons. The sudden movements of re-adjustment send radiations or ether-waves like splashes through space. The longest radiations, used in broadcasting, are made artificially by very powerful discharges of electricity ; the shortest, the gamma rays, used in radio-therapy, are produced in the disintegration of atoms of radium.

If we stand on the shore and watch the waves coming in, we get an impression of one wave following another, and we often speak of the seventh wave as if it had an individuality of its own. But we know that the appearance of the waves chasing one another shorewards is an illusion, for individual drops of water or particles of water do not travel very far. The water has a wavy or undulatory movement with alternate heights and hollows which seem to be coming in our direction as we watch them. Sometimes one wave follows another very quickly, sometimes much more slowly, and so we speak of different wave-lengths, the interval between one height and the next.

Now if we use the words of many (though not all) of those who have studied these things most deeply, we may say that the radiations that are used in broadcasting, that warm the earth, that give us light, that we use to see hidden things (X-rays), are ether-waves, undulatory movements in a universal medium that connects everything. Those who have some doubts as to the need for believing in an "ether" that connects all things, would simply say that space is full of radiations of an electro-magnetic nature.

In any case, whether we speak of the ether or not, there are these powerful wave-movements spreading through space from distant worlds, sometimes, as in the case of the sun, completing their journey in eight minutes ; sometimes, as in the case of the nearest fixed star, taking four years ;

and sometimes, as in the case of distant stars, taking millions of years. There is no harm in repeating these astounding facts which give us hints of the immensity of our universe.

In does not matter much how *far* the radiations travel, whether from a star, or the sun, or a lighthouse, or the electric lamp above our head, in all cases the light-waves are travelling at the same prodigious rate,—186,300 miles per second. But we must not think of the crests and troughs, the heights and hollows being in the line between our eye and the source of the light, as they would be if the light-waves were similar to sound-waves. For the ups and downs, the heights and hollows, of all the radiations we are thinking about—from those used in "wireless" to those used in radio-therapy—are *at right angles* to the direction in which the radiation is spreading, and that means in *every* direction outwards from the source, unless there be some impassable obstacle, such as the heavy shade above our electric lamp.

If we throw a stone into a quiet pond, we see the wavelets spreading beautifully in all directions from the centre where the splash was made. Similarly, the convulsions of atoms in a star make splashes in the ether, or in space, which spread in every direction as far as they can go. Some of them turn back to some extent upon their previous course when they strike on an object that completely *reflects* them. But some of them come to an end of their journey when they enter into a body, which is able to *absorb* them in whole or in part. Since, as we shall notice later on, no power can be lost, the radiant energy that is absorbed is translated into, let us say, the more rapid vibration of the molecules of the absorbing body. The sand feels warm under our bare feet, for radiations from the sun have been transformed into more rapid vibration of the molecules that build up the grains of sand, and more rapid movement of the molecules of air and water that may be entangled amidst the multitudinous particles of sand. Returning to the ripples in the pond for a moment, we see that on a line drawn from our eye to the spot where the stone made a splash, there would be ups and downs, heights and valleys, corresponding to the number of wavelets, and soon, of course, waning away. But in the electro-magnetic radiations, including the light we see, the undulations are *transverse*, that is to say, at right angles to the direction in which the wave at any given point is spreading.

Here it may be noted that sound-waves are not in line with the electro-magnetic waves that we are thinking of just now. Sound-waves are due to undulatory movements of particles in the air or in water. They are not ether-waves. They are massive movements of particles, not electro-magnetic radiations. The noise produced by an explosion some miles away is not long of reaching our ears, but we see the flash an appreciable time before we hear the sound. We might as well compare a snail's pace with the velocity of a bullet.

We have lingered for a little over this question of waves, because the difference between the various kinds of radiation depends on the distance between the crests of the waves. When we speak, for shortness, of "long waves" and "short waves," we mean waves with a long or a short distance (or amplitude) between their crests. Thus the radiations used in broadcasting may have an interval of a mile or more from crest to crest in each wave, whereas the radiations we call visible light have extremely short wave-lengths,—the crests being separated from one another by intervals of very minute fractions of an inch (not more than 0·00008 of a centimetre).

It is plain, then, that we must not picture the waves on the seashore when we are thinking of the electro-magnetic vibrations in space, first because the latter travel in all directions, with undulations at right angles to the direction in which they are spreading, and second because the radiations are all moving at the same velocity, and differ from one another only in the size of their wave-lengths. If there are undulations, some will say, then there must be some *thing* that undulates; and one of the answers to this question is that the radiations are waves in the ether, while the other is that they are discharges of electricity.

One of the great discoveries of modern times is the electro-magnetic theory of light. Newton thought that light meant the inconceivably rapid movement of very minute particles or corpuscles, but that view has in the main given place to the theory that light consists of a number of electro-magnetic radiations with very short wave-lengths, but belonging to the same great series as those with enormous wave-lengths that are used in broadcasting.

What we call visible white light is made up of a blending of various rays which differ in their wave-lengths and produce in our eye and brain the sensations we know as different colours. Suppose we call the range of visible white light *one octave*, from red with the longest wave-length, at the one limit, to violet with the shortest wave-length, at the other limit. Then the whole known gamut of electro-magnetic vibrations appears to include sixty-two octaves. This is a very impressive fact, which we should think over. Let us draw a long double line and divide it into sixty-two divisions, each of which we call an octave (Fig. 3).

Beyond the right-hand end of our figure there are several octaves for radiations with enormous wave-lengths longer than those used in wireless.

Then comes a group of eleven octaves with

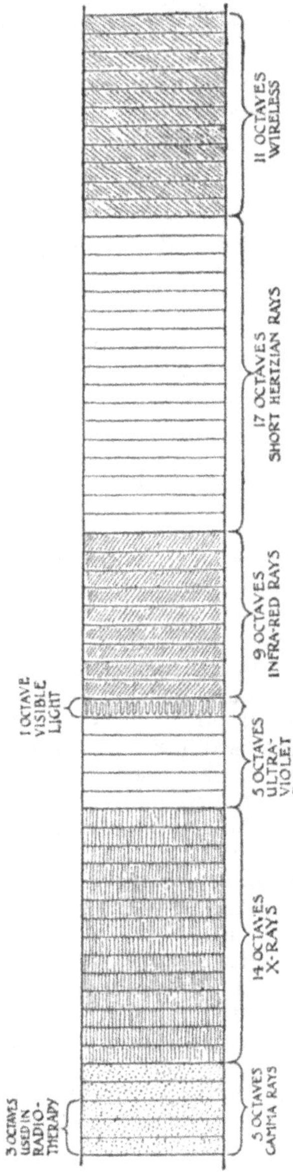

FIG. 3.—THE GAMUT OF ELECTRO-MAGNETIC RADIATIONS.

If visible light be ranked as one octave, sixty-two octaves of electro-magnetic waves have been measured, from those of very long wave-length used in "wireless" to those of very short wave-length used in radio-therapy. Based on a figure by Sir Oliver Lodge. The shading is merely a device to distinguish the various groups more clearly.

very long wave-lengths, the radiations that are used in wireless, though the B.B.C. range includes only a fraction of them.

Next comes a group of seventeen octaves, which are called the short Hertzian waves, after the great discoverer Hertz, who did much to lay the theoretical foundation on which broadcasting is based. All these radiations are produced artificially by very powerful electric discharges.

The next group consists of the infra-red rays which radiate from hot bodies. All the heat rays that come from the sun to the earth are included here, but beyond the limit of those that reach the earth there are others of longer wave-lengths.

A single octave represents in due proportion the visible light which is radiated from hot bodies or emitted by ionised gases, that is to say gases whose atoms have been maimed or convulsed by a radiation.

Of similar origin are the ultra-violet rays invisible to our eyes, but with remarkable effects on bodies both living and not-living. About an octave of these ultra-violet rays is included in the sunlight, but about four octaves consist of ultra-violet rays with shorter wave-lengths, which are generated experimentally.

Another group includes those X-rays which are used in seeing the invisible, such as the bullet buried in the bone, or the pearl in the unopened oyster. They are generated by the sudden stoppage of a fast-moving electron.

The final group includes the three octaves of gamma-rays that are used in radio-therapy, that is to say when the rays in question are used to relieve certain diseases. The gamma-rays include more than another octave, but only those within the three lowest octaves are used in medical treatment. The term X-rays should include fourteen octaves altogether, namely (1) all the "radiology" rays and other gamma-rays at the lower end of the scale, (2) more than half of the octaves of ultra-violet rays towards the upper end of this stretch, and (3) a large range (G) between the two. On a diagram drawn on a small scale it is impossible to reach the precision which is depicted on a gamut where every group of rays is indicated with a high degree of accuracy.

We must not forget a kind of radiation of very short wave-length and great penetrating power that has been particularly studied by Professor Millikan. It belongs to the lower end of the scale. It travels " downwards " from the sky, and is weakened by the amount of air or water it has to traverse. It does not come from the sun, for its strength does not vary with the sun's altitude. Perhaps it comes from the Milky Way, perhaps from diffuse nebulæ in space. It is probably started by the transmutation of one element with another, such as hydrogen into helium. But not much is known as yet in regard to this very penetrating kind of ray.

When we look carefully again at our makeshift diagram we should realise two great facts, first, that light has come into line with the other electro-magnetic radiations throughout the long gamut ; and, second, that the light we see is only one octave out of the sixty-two. Yet it is by means of this one octave that man has come to know all he knows.

A remarkable new idea in regard to energy rewarded Professor Planck's study of the radiation of light-waves from a body heated " red hot." The particles of the heated body are in a state of very rapid vibration, which is sometimes called " material heat," whereas the heat-waves that pass out into space are of the nature of electro-magnetic waves and are sometimes called " ethereal heat." So is it with the light-waves that pass out into space from the heated body that is so hot that it glows. There is a change of energy from the vibrating atoms of matter into the radiations that travel into spaces.

Till recently it has been believed that the light-waves emanating from a heated body formed a continuous stream ; but according to the Quantum Theory the transference of energy takes place not continuously, but step by step. The experiments made by Planck and Einstein led to the conclusion that the radiation or the action takes place in little " jerks." The energy comes out, as it were, in minute parcels or units or quanta. In an ordinary big clock we can watch the large hand moving slowly from minute to minute, but in an electric clock we see a sudden jerk from one position to the next. This comparison may suggest the difference between the old and the new view of radiant energy. The new view is that light, for instance, is not a continuous vibration, but is made up of " atoms," or " parcels," or " quanta " of power. In short, the central idea of the Quantum Theory is *the atomicity of energy!* This is a *very* difficult phrase for a book like this, and yet we can get an elementary idea of what is meant if we picture the difference between a continuous flow of hot water from a tap and an intermittent jerky series of little outbursts.

CHAPTER VII

STATES OF MATTER

IN looking round with inquisitive eyes we are bound to ask: What is this light that our eyes enjoy? What is this warmth that fills us with comfort? This leads to the study of the science of PHYSICS; and when we push our questions into ourselves and ask what seeing means and how the body comes to be warm, we are led to the science

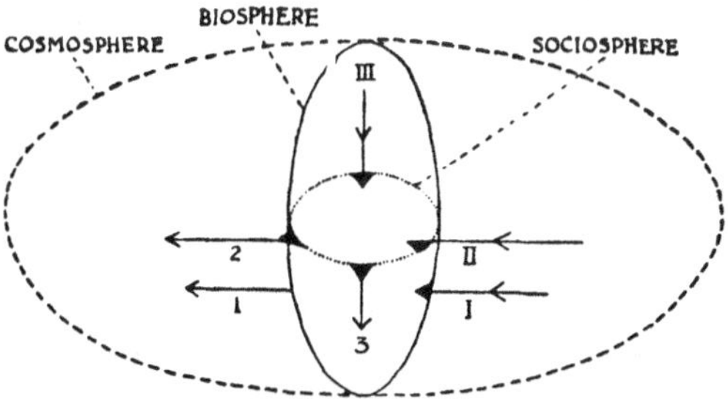

FIG. 4.—THE WORLD OF SCIENCE.

I and II. Influences from inanimate Nature (cosmosphere) may affect animate Nature (biosphere) and human interests (sociosphere).
1 and III. Influences from animate Nature may affect inanimate Nature and human interests.
2 and 3. Influences from man's kingdom may affect inanimate and animate Nature.

of PHYSIOLOGY. Or we may concern ourselves more with matter than with energy, and ask what is the solid earth on which we tread, and what the air we breathe, and what is the nature of the great sea? This kind of question leads us to the science of CHEMISTRY, which is chiefly concerned with the different kinds of matter and with the interactions that often occur between different kinds. One of the first of these questions has to do with the different states of matter—solid, liquid, and gaseous. Everyone knows that dampness in the air means the presence of much water in the form of vapour. It has been drawn up (evaporated) from the surface of the sea or some other body of water. When this water vapour is blown against the cold rocks on the mountain side, it "condenses." That is to say it passes from the vaporous into the liquid state ; it trickles down the rocks to form one of the innumerable runlets that feed the springs or the streams. But if great cold should set in, the liquid water becomes a solid icicle. Thus we are all very familiar with water as vapour, as liquid, and as solid. The steam that escapes from an engine or from our nostrils on a frosty morning consists of molecules of water in a gaseous state, for the distinction between vapour and gas is quite unimportant. A vapour is the gaseous state of a substance that is in ordinary conditions a liquid or a solid. The first point in our present study is simply that what is familiarly true of water is true of many other substances, that they occur in three states—gaseous, liquid, and solid. Even the air may be made into a liquid, and carbonic acid gas can be forced into a solid like snow. What we must try to picture is the differences between these three great states—gas, liquid, and solid.

When a strong beam of light enters a darkened room we see a great crowd of motes or dust particles dancing as if they were alive. In the strict sense we do not really see the particles, for they are too small to be seen ; and in ordinary diffuse light they *are* invisible, as we very well know. What we see are minute haloes formed by reflection from the surfaces of the invisible particles. Some of the rays of the beam of light rebound from the particles and form little discs of light, much larger than the particles themselves.

It is interesting to sit and watch the dancing motes, partly because we had no suspicion that there was so much dust in the air of the room ; partly because we are made to think of the draughts and temperature-differences that make the particles swirl about ; and partly because the motes give us a rough picture of the incessant movements among the molecules of any and every gas.

The molecules of a gas are in rapid perpetual motion, in

all directions, in straight lines, and no one molecule gets very far without colliding with another. On an average a molecule travels about twenty times its own length between two collisions, but that means only a millionth of a centimetre (a centimetre being about 2/5 of an inch). Later on, in studies beyond this book, an attempt should be made to understand how investigators found out about these collisions. In the meantime they must be taken on trust; and that is why we began with the dancing motes which would have been invisible had it not been for the beam of light.

It has also been discovered that the molecules of a gas are travelling with great velocity,—at the rate of about a quarter of a mile in a second at ordinary room temperature. In a furnace they are travelling much more rapidly, and in the sun, as we have already mentioned, the rate may be a hundred miles in a second. This rapid rushing of particles is what is called "material heat," to be distinguished from the "ethereal heat" which consists of waves of energy travelling through space. But keeping to our room, we say that the molecules of oxygen and nitrogen and carbon dioxide are travelling rather faster than sound travels. Their rate is about a quarter of a mile in a second. In this distance (quarter of a mile) and during this time (one second) a molecule suffers 5000 million collisions. This would be intolerable in a human dance! It might be thought that the collisions would slow down the molecules, but it is part of the theory of gases—a theory that works so extraordinarily well that it must be true—that the molecules are perfectly elastic. That is to say, they rebound from one another with no loss of speed and with no loss of heat or of any other form of energy.

A toy balloon is filled with gas and its wall stands out tense and firm. Using scientific words, we say that this is due to the pressure of the gas, which just means the continual impact of the flying molecules. They are continually bombarding the internal surface of the walls of the balloon; and if the temperature is raised, the molecules travel more rapidly, and there are more impacts per second. Getting into a very warm room, the toy balloon may suddenly burst, which means that the bombardment of molecules is too much for the walls to stand.

If one fumbles in lighting a gas-jet, and allows some of the coal-gas to escape into the room, the molecules spread themselves everywhere as far as they can go. Soon they will be found in every corner of the room, but so few and far between that we cannot smell them. In this case the gas is in a very diffuse state, differing from the gas in a dense state in having far fewer molecules in a given space and far fewer collisions between the molecules; but the molecules of the coal-gas in the room will be colliding, of course, with the molecules of the mixture of gases that form the air. Our atmosphere seems thin enough, but there are stars that have a mean density a thousand times less, and thinner still is the matter in the vast spaces through which light travels on its way from the stars. Until it comes near the earth the light is only passing one atom per cubic inch throughout its long journey.

But now instead of thinking of molecules rushing apart, let us think of them coming close together.

When the volume of a strong vessel containing gas is reduced, by forcing in the lid, or in some other way, the flying molecules will be crowded, and there will be more impacts on the walls. As we say in three words, *the pressure increases*. If we can continue the compression, the molecules become so crowded that they are attracted to one another, as a stone to the earth, and they tend to cluster in little groups. They form a close-packed film on the walls of the vessel. In short, the gas condenses into a liquid. In practice, this is made easier by cooling the vessel to a very low temperature, for this slows down the movements of the molecules. There was an interesting period in the last quarter of the nineteenth century when some obstinate gases like oxygen and nitrogen were liquefied. This kind of experiment, begun by Faraday in the Royal Institution in 1823, was crowned in the same building in 1898, by Sir James Dewar's liquefaction of hydrogen.

The molecules of a liquid are still rushing to and fro, but they are much more crowded and have much less freedom of movement. They are so near one another that it is difficult for them to get nearer unless they are subjected to great pressure or to great cold, or to both combined. Yet every one knows how those on the surface of water tend to break free or evaporate into the air; while it is also a familiar, though striking, fact that those on the surface may hold so tightly together that they form a " skin " on which a dry needle may be floated.

When molecules are forced to come together still more closely than in the liquid state, the result is a solid. They may form a higgledy-piggledy crowd, so dense that individual shifting is scarcely possible. This is the case in glass, and it is a familiar fact that no one can bend a window pane more than a very little. It is possible, however, to send a helium particle right through a pane of glass. The particle is so very minute and it may have such a high velocity (even 18,000 miles per second) that it can get through before the glass has time to break!

But the crowded molecules may assume an orderly arrangement, like a regiment, not like a mob, and this spells "crystal." Just as some liquids are much more fluid than others—water more than syrup—which means that the molecules have different degrees of free play, so there are varieties of solidity. The lump of cobbler's wax is much more plastic than the lump of lead. In no case, however, should we think of the molecules of a solid as quite motionless. For while they cannot travel or collide, they are in a state of vibration within a narrow range, and each kind of solid has its own favourite frequency of vibration. Let us recall the familiar fact that heating the solid increases the rapidity of the molecular vibration, until they shake themselves loose from one another, and the solid melts. It may also be noticed that the particles on the surface of some solids are always liberating themselves and passing off as vapour in the air. This "sublimation" from a solid is obviously the same sort of phenomenon as "evaporation" from a liquid, and it may be well illustrated by the camphor ball that has been placed in the museum case to warn off insects. It gives off molecules for years till some day we notice that it is no longer there!

What we have said in this study is only a beginning; but it is a great gain if we are intellectually humble enough to think of gaseous, liquid, and solid states of matter as three degrees in the freedom of movement among the constituent molecules. In a gas there is a wild dance with incessant collisions; in a liquid there is surging to and fro, but much less freedom; in a solid there is hardly more than what might be compared to the almost imperceptible swaying seen in a body of soldiers standing at ease.

Before leaving the subject, only to return to it, we hope, a little must be said in regard to solutions, colloids, and films.

When we put a cube of lump-sugar into our tea, it disappears, passing into solution. What exactly happens is by no means so simple as it looks. It is plain, however, that the molecules of the sugar, which were fixed together in crystals, are now moving about in the solvent that we call tea; and to avoid complications we may as well think simply of hot water. After a little stirring every drop of the water is as sweet as any other drop, so the solution is a *homogeneous* mixture of one substance (sugar) with another substance (water). Of all the solvents water is the chief; that is to say, more substances are soluble in water than in anything else.

But some of the molecules in the liquid water (H_2O) break up or are dissociated into hydrogen ions (H) and hydroxyl ions (OH—); and a further complication is that water seems to help in dissociating or ionising the molecules of substances that may be dissolved in it. When a substance is thus broken into imperfect or maimed molecules it is called an electrolyte, which means that it will conduct an electric current. This is not the case with our sugar solution or with a starch solution, neither of these substances being an electrolyte, but it is the case with a salt solution. It is a well-established fact than when common salt (sodium chloride) is dissolved in water, most of the molecules break up into two parts, an *ion* of sodium and an *ion* of chlorine. The sodium ion carries a charge of positive electricity, the chlorine ion an equal negative charge. These electrically-charged ions give the salt solution its power to conduct electricity, which the sugar solution has not got.

Of great importance is the colloidal state exhibited by many kinds of matter, such as living matter itself. In substances like raw white of egg and liquid gelatine, to mention two typical colloids, there are innumerable groups of molecules dispersed in a watery medium with which they do not mix. Cod-liver oil emulsion is a good example of innumerable droplets "dispersed" in a medium; but in colloidal gold the dispersed particles are solid, and so it is in hundreds of cases. When the amount of the liquid medium is small in proportion to the included substance, so that the whole has a certain solidity, we speak of a *gel* not a *sol*, as when liquid gelatine "sets." Many of the changes in vital activities are associated with alternations of *sol* and *gel* states.

It is not difficult to *begin* to understand why the colloidal state of matter is so effective. Its potency depends on the relatively large surface presented by the countless particles and droplets, for chemical and physical changes take place on these surfaces. We may picture an archipelago with thousands of islands ; there is a vast length of coast-line on which trading may go on. The particles in a colloid are like the islands in that sea.

Finally, it seems almost necessary to speak of a fourth state of matter—the film, such as we know in a soap-bubble or in the skin of oil that calms troubled waters. All the physical properties of a substance change when it is in the film state, and in everyday life much depends on the control that the filmy cell-membrane exerts on the passage of substances in and out of the cells that build up all but the simplest living creatures.

CHAPTER VIII

THE STRUCTURE OF MATTER

FROM the sun, which is our great fountain of power or energy, we passed naturally to the earth and began to take stock. We got glimpses of its changes from age to age—changes in the arrangements of its matter and energy. We lingered for a little over the different forms of energy, which may be defined as the power of doing work; and then we passed to the different forms that matter, such as water, may take: gaseous, liquid, solid, and filmy. But we must not make too much of the difference between these two great realities MATTER and ENERGY, which used to seem so different, for modern investigators of Physics have shown that matter is made up of different combinations of electrons and protons,—that is to say, negative and positive charges of electricity, and electricity is a form of energy.

We must repeat that we are not forgetting the greatest of all the realities, namely Mind; but we shall come later in our studies to that quite different kind of power. We *cannot* forget it, for it is, of course, by our mind that we know about other powers.

It is a pleasant idleness to let the fine sand trickle through our fingers as we sit by the shore. When we know enough about it and peer into the minute fragments, we seem to be skimming the pages of a romance, for each grain of sand has a history. Here is a minute crystal of quartz and there is a rounded granule of black basalt. At one moment our eye catches a miniature garnet, at another a speck of manganese-ore. Then again there are tiny flakes which we recognise as fragments of a sea-urchin's shell. Some quick eyes can catch the flinty spicules of a sponge.

The sand is very fine, but we know that if we brayed it with a pestle in a mortar, we could make it finer still, and it would be difficult to keep it from being blown away by gusts of wind. In the neighbourhood of mines where they crush the rocks in the search for gold, the air is apt to be full of very minute sharp-edged fragments of quartz, which are drawn into men's lungs, and often do much harm when they lodge in the narrow passages and cannot be breathed out again.

In various ways besides by pounding, we can get *very* minute particles; and our question is: What should we come to if we continued the theoretical pounding further and further, even beyond the limit at which any particle is visible by itself. The answer is that we should reach the invisible MOLECULES, which are defined as the smallest particles of a substance that can exist in a free state. Thus the molecules of oxygen in the air are not the smallest amounts of oxygen that can take part in a chemical reaction (these are the ATOMS), but the smallest particles that can exist by themselves in a free state, that is, neither in combination with other molecules or with atoms, nor yet in solution. The molecules of oxygen in the air are believed to consist of two atoms of oxygen in close linkage (O—O or O_2). The molecules of carbon dioxide in the air are thought of as composed of an atom of carbon linked to two atoms of oxygen (O=C=O, or CO_2). The water-vapour in the air is made up of molecules in which two atoms of hydrogen are bound to an atom of oxygen (H—O—H, or H_2O). Yet the helium molecules, which occur very sparsely in the atmosphere, are believed to consist of single atoms. In other words, atom and molecule *are sometimes the same*, though in most cases a molecule is made up of several atoms. Thus the molecule of mercury is believed to consist of one atom of mercury; the molecule of oxygen is pictured as two atoms of oxygen; and the molecule of water as two atoms of hydrogen linked to one atom of oxygen. A *chemical element* is any particular kind of homogeneous matter built up of similar atoms, and *an atom* may be defined as the smallest particle of a chemical element that can enter into or be expelled from a chemical combination.

Once more, if we could continue dividing and dividing grains of common salt, sodium chloride (NaCl), into smaller and smaller particles, we should in theory reach invisible minutiæ, the breaking of which would yield not smaller particles of salt, but infinitesimal portions of the metal sodium and of the gas chlorine. Thus we may add a clause to the definition of molecules, which would then read: the smallest particles of a substance that can exist in a free state, yet having the same composition as any larger mass of the substance. A molecule of salt is as truly salt as a grain or a spoonful.

Visible particles, then, are made up of invisible molecules, and a molecule is built up of atoms, except in cases where the molecule consists of a single atom. Having cleared the ground, let us now turn to the atom.

The word atom means "what cannot be cut," and part of the old idea of an atom was that it could not be divided into anything smaller. According to the modern view, an atom may have a complicated structure! It may be like a little solar system, like a constellation of stars! But if a molecule is invisible, and an atom still more so, how can we know anything about its structure? The answer is that the structure of the atom is an ingenious invention of mathematicians and physicists which must be something like "the real thing," inasmuch as the properties of matter are such as they would be if the invented structure were real. The invention must be a good guess at truth, for several prophecies based on it have been remarkably fulfilled.

The simplest of all the atoms is that of hydrogen, which is believed to consist of a hydrogen nucleus or PROTON, carrying a charge of positive electricity, and of a single ELECTRON, carrying a unit charge of negative electricity, and revolving round the nucleus at a high velocity. So it is easy, so far, to picture the outline structure of a hydrogen atom—a nucleus of hydrogen and an electron revolving around it, like the moon around the earth. But this hydrogen atom is far too simple to be a good sample, and we may take in contrast the atom that is the most complex as yet known, that of the element called uranium. It seems to have (1) a core of 238 hydrogen nuclei or protons; and (2) along with these, 146 inner electrons; and (3) outside these again, 92 outer electrons. Thus if we think of the centre of the uranium atom (the core and the inner electrons) as a sun, we may think of the 92 outer electrons as so many planets. The poet Blake spoke of seeing "a world in a grain of sand"; and that is what modern science has verified; the atom of uranium is like a constellation. For the time being (1928), that is a picture that commends itself to many of the acutest living minds, but even as we write the picture is changing in detail. In any case the great conclusion stands out, that all the different kinds of matter differ from one another only in the number and arrangement of their electrons and protons. When we think of carbon, hydrogen, oxygen, nitrogen, sulphur, phosphorus, chlorine, mercury, iron, silver, gold, radium, uranium, and all the rest of the chemical elements, we must think of them as depending on different arrangements of electrons and protons, the latter being otherwise called hydrogen nuclei. Matter seems to dissolve into electricity. The electrons and protons are the building stones of the universe, " the stuff out of which worlds have been spun."

Here then is one of the greatest unifications that science has ever made : the different kinds of matter differ only in the number and arrangements of their electrons and protons. If the atoms were not such " fierce little worlds " with great internal activity, we might say that they all show the same stones and mortar, but with different architectural styles.

Some of the atoms, especially the heavy ones at the top of the series, are unstable and as it were overcrowded. " They are apt to explode," as Sir Oliver Lodge puts it, " flinging away every now and then a superfluous electron and proton, thereby losing their original chemical character and becoming a different element. After radium has flung away five such electric charges of each kind, it is indistinguishable from lead. But it is very leisurely about it. Centuries may elapse between successive explosions of neighbouring atoms."

Several series of chemical transmutations—somewhat like pedigrees among changeful plants and animals—are now known, but the atoms must be allowed their own way. They cannot be coerced as yet into obedience to man's purposes, and the changes that occur *on the earth at least* are all in the same direction. There is breaking down, but no building up ; there are disintegrations, but no integrations. To put it in simpler language, not perhaps clearer, the clocks are all running down ; and at present there seems to be no winding up. But there must have been winding up in bygone ages, to yield complex atoms like uranium. Perhaps there is winding up now going on in sun and star.

No doubt the clever synthetic chemist is continually

building up complex carbon-compounds, artificially "synthetising" sugar, alcohol, salicylic acid, adrenalin, and all sorts of perfumes and pigments, drugs and dyes, explosives and fertilisers. But apart from the fact that this is "artificial" not "natural," with man's intelligent finger in the pie, we must notice that the creative chemist is working in the domain of *molecules*, and this is very different from building up an *atom*.

Our point is that as things are at present on the earth, complex "radio-active" atoms are known to break down, but we have no knowledge of their being built up. Lead can be born, but uranium, thorium, actinium, radium, and their relatives of high atomic complexity seem only to die. Uranium may give rise to protactinium, which produces actinium, which produces lead. Or uranium may give rise to ionium, which may give rise to radium, which, by giving off helium, may produce lead. And thorium may also give rise to lead. Thus, without going any further, we are confronted with a very puzzling fact that there are several different kinds of lead—actinium-lead, radium-lead, and thorium-lead, all the same in their chemical reactions and yet slightly different in internal structure. They form what are called "isotopes," and there are more than we have mentioned, besides the mixture called "ordinary lead," perhaps eight altogether. Isotopes are elements that are chemically the same, yet different in the internal make-up of their atoms.

An interesting use is made of the sinking down of uranium or the like into lead; it is used as one of the ways of discovering the age of the earth. If lead is found in a rock along with uranium, the natural inference is that the lead is the outcome of the slow but sure disintegration of the uranium. But the rate at which the transmutation takes place is known experimentally. Thus the question comes to be: How long would it take the uranium to form the proportion of lead that the mineral shows. This kind of argument leads geologists to conclude that the older rocks must be far over a thousand million years old. How much older must the sun be!

Nothing in the study of atoms has been more striking than Sir Ernest Rutherford's experiments on bombarding their nuclei or cores. We quote a passage describing this from Sir Oliver Lodge's book on "Atoms and Rays."

"The nuclei could not be shattered, or got at in any way, by any such trivialities as high temperature, extreme cold, enormous pressures, chemical explosions, or anything of that kind. They were far beyond the reach of these trifling perturbations. But the projectiles fired off by radium at a speed of several thousand miles a second, were not so insignificant. And Rutherford arranged to bombard the nucleus of any desired atom by means of these projectiles. The nuclei were targets excessively difficult to hit, because they were so ultra-minute; and thousands of shots might go by them without achieving anything. But then, hundreds of thousands of shots were available, any number in fact; so that sooner or later there was bound to be a hit. And then something happened. Briefly, the nucleus broke up, and hydrogen flew out. This is one of the most remarkable experiments ever made by man." How a nineteenth century expert would have smiled if he had been told that hydrogen could be knocked out of an atom of, say, nitrogen.

Can we say anything in regard to the dimensions of atoms? The smallest particles visible with the unaided eye are about one-tenth of a millimetre in diameter, a millimetre being about one twenty-fifth of an inch. The microscope reveals particles one hundred times smaller than we can see without it, but no matter how powerful the microscope, we can never get it to make visible any particle that is smaller than half the wave-length of the light by which it is illuminated. The bodies that are seen in an ultra-microscope are the haloes of light sent off from the surface of quite invisible particles.

Let us repeat, with our own eye-lenses we can see a particle about one-tenth of a millimetre in diameter. With the lenses of a high-power microscope we can see a particle one hundred times smaller. Two hundred times smaller still are the large molecules of starch, whose presence gives, not a milkiness, but a slight opalescence to a water solution. An atom of hydrogen is one hundred times smaller than these; and the atom of hydrogen consists of one proton with a single electron revolving at a relatively great distance from it. This is the world of the infinitely small which has been and is being revealed as the result of Becquerel's discovery of the rays of heat and light that stream forth from radio-active substances. This is a fine instance of the way in which a particular discovery—radio-activity in this case—may give man a new world. This is what science is always doing—making the world new; and it has become so new within this century that those who are getting old have to rub their eyes very hard. And we cannot suppose that this making of new worlds is going to stop, or that the "electron," for instance, is one of the last words of science. Thus we notice that Sir J. J. Thomson, to whose genius the twentieth century picture of the structure of matter is largely due, has been giving a lecture entitled "Beyond the Electron," in which it is explained that the electron is a system built up of many smaller parts.

CHAPTER IX

THE CONSERVATION OF MATTER AND ENERGY

CHEMISTRY has mostly to do with transformations of matter, and Physics with transformations of energy; and the foundation-stone of them both is that in any ordinary experiment that we make, or operation that we arrange, the sum of matter and of energy is the same at the end as it was at the beginning. This is called the conservation of matter and energy.

We light a candle and it burns away until nothing visible remains. We put a solid block of coal on the fire and it burns away until nothing is left save some smoke in the sky, some soot in the chimney, and some ashes on the hearth. We put a big lump of sugar into our tea, and it soon dissolves away; but when we taste the tea we know that the sugar is still there, though it has entirely changed its form. When we think over the dissolving sugar, we may begin to suspect that in the other cases also matter only seems to disappear; it merely changes from one form to another. If we could collect all the carbonic acid gas and other gases, including the water-vapour, given off by the burning candle, we should find that they weighed just the same as the unlighted candle *plus* the oxygen that was used up from the air in the process of burning. There is not much to show for the barrel of gunpowder after it has exploded, except that it may have done untold mischief, yet if we could collect all the gases, we should find that the total amount of matter is the same at the end as it was at the beginning. This is what is meant by the CONSERVATION OF MATTER. As the great chemist Ostwald has put it: "the total mass of the substances taking part in any chemical process remains constant."

As the masses of bodies—the quantities of matter in them—are at any one place proportional to their weights, the conclusion may be read, that in any ordinary chemical process, the *weight* remains constant. Indeed if there were any discrepancy between the weight of the material at the beginning and the weight of everything at the end of the operation, the chemist would at once conclude that something had gone wrong, or that something was taking place that he did not understand. It was by interesting himself in a little discrepancy in weight that Lord Rayleigh, in 1892, got on the track of Argon, a very sluggish dense gas occurring along with nitrogen in the air. Lord Rayleigh noticed that nitrogen obtained chemically from some compound like ammonia or saltpetre was about one-half of 1 per cent. lighter than the nitrogen obtained from the atmosphere, for instance by passing air through a red-hot tube packed with copper filings. The minute difference in weight was found to be due to the fact that the nitrogen from the air had mixed with it a previously unknown heavier gas, Argon. The same discovery was made independently by Sir William Ramsay, and the two distinguished investigators published a joint memoir on the new gas.

The laying of the chemical foundation-stone which we call *The Conservation of Matter* cannot be put to the credit of any one man, but it should always be particularly associated with the great French chemist Lavoisier, who was beheaded (1794) at the time of the French Revolution by the misguided "Reds," who shouted "the Republic has no need of savants." It is often and truly said that "science begins with measurement," and by making a more accurate balance than had been previously devised, Lavoisier was able to show that no matter ever goes amissing. To take one instance, he passed water-vapour over red-hot iron turnings, with the result that the water broke up and formed the gas hydrogen. Lavoisier weighed all the materials before the operation and all the materials after the operation, and found that the weight was the same.

In what we have said we have referred to all ordinary chemical operations, and laid emphasis on the experimental conclusion that in such cases no destruction nor creation of matter ever takes place. But this is not to say that there may not be a breaking down of atoms, as in radio-activity;

or that matter may not have been formed long ago from something that was not matter, namely energy. But here we get into waters too deep for us just now.

Conservation of Energy

For everyday purposes we may think of Energy as the power of doing work. The swiftly flowing river does work in turning the mill-wheel, and the wind does work in driving the sailing ships across the sea. The expanding steam in the cylinder of the engine does work in moving the piston, so that the wheels go round; and the fire does work in tearing asunder the molecules of cellulose that build up the log of wood. Electricity does work when it drives the tramway-car, or makes the ventilating wheel in a room spin round at a great rate. Part of the sunlight does work when it enables the green plant to build up sugar out of carbonic acid gas and water. The explosive does work when it breaks the rocks asunder and sends great stones hurtling through the air. Our many muscles do almost silent work when they contract and pull one bone nearer another, levering us along the ground when we walk. Our food also does work in forming the fuel for the quiet engines of the body. Of course "work" is often a subtler thing than moving a visible body like a wheel or a ship or a stone, but, to begin with, we do not go far wrong if we think of energy as "the power of doing work or mischief."

Energy is always changing from form to form, but none is ever lost. The energy of the waterfall is changed into electricity, which may be changed into light in a distant city. Like the mythical Greek personage called Proteus, or like the génie in "The Arabian Nights," energy can change rapidly from one form into another. But when we think of energy being transformed from one guise to another, we soon get into difficulties unless we recognise that "the power of doing work" need not necessarily be in active operation. The mill-pond on the slope is fed by a small stream, whose water is energetically pulled down by gravity nearer to the centre of the earth. But the mill-pond's accumulated energy is not in operation until the sluice is opened and the water rushes down to move the mill-wheel. This energy not in action is called Potential Energy, in contrast to Kinetic Energy, which is "at work."

A complex chemical substance like gunpowder has a great store of potential energy, but it cannot do anything until the trigger is pulled by means of a spark, or in some such way. But whenever the trigger is pulled, the surrounding oxygen rushes into union with the explosive and there is a violent tearing asunder of molecules. So much gas is suddenly produced that great weights can be lifted and much work or mischief done. And apart from movement on a big scale, there is also a transformation of part of the explosive's potential energy into the kinetic form of heat and light. But the very opposite kind of change is seen when the kinetic energy of part of the sunlight is used in the green leaf to help to build up sugar and other carbon-compounds, a form of potential energy which may become the source of motor power to an engine, an animal, or a man. Or, to take an example from man's device, the motor energy of the waterfall may be transformed into the radiant energy of powerful electric discharges, and these in passing through the air may form the potential energy of nitric acid. This is still further compounded, and the results, such as ammonium sulphate, are used as fertilisers, which become the food of plants. By further upbuilding in the plant there is a formation of the nitrogenous carbon-compounds, called proteins, like the gluten of wheat, which have still more potential energy than the fertilisers had. By eating the plant-proteins the horses gain energy to plough the field. Thus there is a continual transformation of energy from kinetic form to potential form, and from potential to kinetic. But no power is ever lost.

Yet this sounds too good to be true, and we hasten to add the saving clause which Lord Kelvin first made clear, that every transformation involves some practical loss. For some fraction of the energy involved in the operation passes into the form of diffuse heat, which can hardly be got hold of again. Every transformation involves a certain amount of "degradation," meaning by degraded energy that it is less capable of being transformed than it was before. In some cases the "degradation" is very subtle; thus the fermentation that takes place in a fire-fly's luminous organ gives rise to cold light without any accompaniment of heat-rays. As far as the production of light-energy is concerned, the fire-fly is almost uniquely economical, for in man's successes in light-production there is always a fraction of loss in the simultaneous production of heat. We suppose that it would be theoretically possible to recover the energy of the fire-flies' lamps, but the heat-rays that radiate into space are for the most part irrecoverable. Thus in ordinary conversation we refer to " beating the air " or " heating the air " when we wish to hint at hopelessly wasteful proceedings.

Another large fact that must be clearly understood is that almost all the energy that is usable on the earth comes from the sun. The moon helps a little in causing the tides, which are sometimes used as a source of power, but the fountain and origin of almost all available terrestrial energy is the sun. The river that turns the mill-wheels is fed by the clouds, consisting of water-vapour, which the warmth of the sunshine has lifted from the ocean. The wind that drives the wind-mill and bears the ships across the sea is due to the sun's unequal heating of earth and air. The railway locomotive that carries man and his products from one end of the country to another is burning the coal which is the buried energy of the sunlight that penetrated the flowerless forests of the Carboniferous Age. The oil-engines of the steamship are also driven by the transformed sunlight of distant ages. The energy of our galloping horse and the energy of our hard-worked heart are alike due to the food that is eaten, and sooner or later this must be traced back to the chemical processes that go on in the green leaves in the sunshine.

No doubt the streams of heat and light that pour forth without ceasing from a disintegrating radio-active substance are not traceable to the sun in any direct way, but these radiant energies are not to any appreciable extent available for man's operations. On the other hand, as we have already stated, there seems to be general agreement that the sun's generous scattering of energy is sustained by a continual breaking down of atoms, in its unthinkably hot furnace. And similarly for the twinkling light of the stars. If man should learn to break up atoms, as radio-active atoms spontaneously break up, then he would have at his disposal, for good or ill, inexhaustible supplies of power. He would be independent of the sun! But the thought of the powers that are hidden in the dust is somewhat overwhelming.

Let us try to imagine a hermetically sealed little world,

not parting with any energy to the outside nor receiving any energy from the outside, yet busy for a while with all sorts of chemical and physical changes, the sum of all the matter and energy in that little insulated world would remain constant. This is the idea of the CONSERVATION OF ENERGY. In Clerk Maxwell's words: "The total energy of any material system is a quantity which can neither be increased nor diminished by any action between the parts of the system, though it may be transformed into any of the forms of which energy is susceptible."

Let us return for a moment to our imaginary closed room, from which there is no leakage of energy, and into which there is no seepage. If we suppose that we can estimate exhaustively all the energies that are in the room to begin with, then, after a whole series of operations has taken place, the amount of energy at the end will be the same as at the beginning. What happens may be compared to what occurs in a change-office outside the gates of a great Exhibition or the like; there are thousands of transactions in the course of the day, but the cash at the end should be precisely the same as it was at the beginning. But whereas the change-office would open in the morning with much silver, and would end at night with many banknotes, a very condensed form of power, the tendency in Nature, as we have said, is in the opposite direction—towards the pence of diffuse heat. This is the general idea of the Conservation of Energy, but we must not be in a hurry to say that what holds for a closed system in an experiment is true for the world as a whole.

CHAPTER X

THE BEGINNING AND BASIS OF LIFE

SO far we have been thinking of not-living things and powers, now we pass to living creatures or organisms. They consist, it is true, of matter, and their life involves continual transformations of energy, so that there is a chemistry and a physics of the living body, whether of plant or animal or man. Yet their activities are in many ways different from those of rivers and tides, planets and stars ; they are *individuals* that feed and grow and multiply, that wind themselves up when they run down, that struggle and get things done, that often change a little from generation to generation. Sometimes it is quite plain, as in horse and dog, that they have a mind of their own. Thus a special science is required—the science of Biology, which has various subdivisions, such as Botany and Zoology.

For untold ages the earth was void of life. The crust was too hot, and no kind of living creature that we can think about can exist where there is not water in a liquid state. For living matter always has at least 75 per cent. of water in its composition. Gradually, however, as suggested in previous studies, the crust of the earth cooled, water-vapour condensed to form lakes and seas, and the sunshine broke through the thick canopy of cloud. The earth became fit to be a home of life.

How living creatures began to be upon the earth, no one knows. It may be quite true to say that a living creature is " a handful of dust which God enchants," but this is a religious way of *interpreting* what happened, it is not a scientific *description*. While it must be admitted that a convincing scientific description is still awanting, the favourite suggestion in scientific circles is that living creatures of a very simple sort emerged in some secret spontaneous way from non-living materials. It must be admitted that there is not any hint of this happening nowadays, but this does not warrant us in saying that it may not have happened long, long ago. When living creatures make an unexpected appearance to-day, like maggots in a dead bird, like threadworms in a saucer of decaying paste, it can always be shown that they or their very young stages got in from outside in some way or other. The mother blow-fly laid its eggs in the dead bird; it may be that a gust of air carried the developing embryos of the threadworm on to the saucer of decaying paste. In every case that has been carefully studied there is a verification of the conclusion "omne vivum e vivo," all life from life. Even more precisely than that, every living creature arises from another living creature *of the same kind*, though sometimes there is a noteworthy new departure of variation, as when a long-haired rabbit appears in a short-haired race, or a cat without a tail in a normal tail-possessing lineage.

We have confessed that no one can tell as yet how living creatures began to be upon the earth, or in the waters under the earth. In the scientific sense of the word "knowing," we do not know how the breath of life began. Yet, as it is very unlikely that germs of life came to the earth from elsewhere, borne in the crevices of a meteorite or wafted through space in the cosmic dust, we are brought more and more to consider the possibility that the first organisms may have arisen from non-living matter,—from specks of nitrogenous carbonaceous jelly naturally built up in some quiet pool in the light of the sun.

The experiments of Baly and others seem to show that light shining on water containing carbonic acid gas may bring about their union, the result being a very simple carbon-compound called formaldehyde, with the formula CH_2O. From that starting-point it is possible, with the help of more light, to build up sugar; and we know that this is always being formed by green leaves in the sunshine. The chemist has learned to mimic what takes place in the living leaf.

When there is a thunderstorm the electric discharges may bring about in the damp air a union of nitrogen with hydrogen and oxygen, and there may be a production of nitrite of ammonia, which the rain may bring down to the soil and to the pools. It may be recalled in passing that man has now learned to capture the free nitrogen of the air by using very powerful electric discharges, thus forming nitric acid which he utilises in manufacturing fertilisers. We have already noted that these fertilisers become food for plants, such as the wheat in the fields, so it is not too much to say that man is now using the thin air in the making of bread!

Let us suppose that the rain brought down ammonium nitrite into a sunlit pool where carbon dioxide and water were uniting to make formaldehyde and other simple carbon-compounds, these might capture the nitrite and form "amino-acids," which have been called the "basic substances" of life. For amino-acids in combination form proteins, like white of egg and the casein of milk,—proteins, which are the main constituents of all living matter. Our point is that in the pool there might be in favourable circumstances long ago a beginning of living creatures of a very simple kind. No one is really sure, but there is nothing preposterous in the idea that living creatures emerged by a continuation of the natural processes of a previously life-less, but not mind-less, world. As we cannot juggle "mind" out of "matter," we must suppose that some form of mind was there all the time. "In the beginning was Mind," we repeat to ourselves when we try to think things out.

The first living creatures probably began in the sea, though it is possible that they began in fresh water. If we consider the humbler animals that are living to-day,—namely, the simplest single-celled animals, the many-celled sponges, the stinging animals like sea-anemones and jelly-fishes, the multitude of worms, and so on, we find that the *great majority* are marine, a small minority occur in fresh water, a few simple single-celled animals are found in the damp soil—forerunners of the earthworms who followed their example long afterwards. If we consider, for instance, the class of sponges, including many hundreds of different kinds, we find that all live in the sea except the members of one family of fresh-water sponges or Spongillids. Similarly, among the Stinging Animals (Cœlentera) there are a few fresh-water polyps and beautiful fresh-water swimming bells or medusoids, but all the others are marine—thousands of different kinds. Here then is a strong argument that the original home of animals was in the sea.

A living creature cannot make energy any more than an

engine can. Both are merely *transformers* of matter and energy. But the living creature has this unique secret, that it persists for a longer or shorter time as a going concern. It requires food just as the engine requires fuel, but it has processes of up-building which counterbalance for a time the processes of down-breaking. It can wind up its own clock. For a time it can balance its accounts, in a way that no engine can do. The first living creatures were probably minute specks that floated in the sea, neither quite plants nor quite animals, but what are called Protists. They were able to utilise the energy of the sunlight to build up carbon dioxide and water into sugars and other carbon-compounds, which formed the fuel of the living fire. The breaking down or combustion of the carbon-compounds was the source of the energy which was expended in moving and growing. So we start with a sunlit sea teeming with invisible Protists, from which in the course of time there evolved both plants and animals.

Let us now change the subject a little and think of the material out of which all living creatures are built up, the protoplasm that Huxley defined as "the physical basis of life." While it is quite true that there is "one kind of flesh of men, another flesh of beasts, another of fishes, and another of birds," Huxley laid emphasis on the fact that all kinds of "flesh" have a similar physical basis. Protoplasm or genuine living matter shows itself as a clear liquid in which there are suspended multitudinous minute particles, or unmixing droplets, occasionally, especially in plant cells, in a state of constant ("Brownian") movement. When the protoplasm dies the liquid "sets" as a jelly, and the Brownian movement of the particles ceases. A similar change from "sol" to "gel" often takes place temporarily during life. Every one is familiar with the coagulation of white of egg in boiling water, or with the liquefaction of gelatine in similar conditions; but the changes in protoplasm are naturally somewhat subtler. Raw white of egg can be partly solidified at a low temperature and entirely solidified by drying; and it is possible to work the change the other way by raising the temperature again or by adding water. But protoplasm can pass from the solid to the liquid, from the gel to the sol, or *vice versa,* without any change in the temperature or in the water-content. This is part of its secret.

Before saying more about the physical characters of protoplasm, let us think of it chemically. Living matter cannot be analysed as such, for the methods of analysis kill it; and death may mean that large molecules rapidly tumble down into smaller ones. But the chemist tells us that there are no elements in the protoplasm that are not common enough in the surrounding non-living world. Those that are invariably present are carbon, hydrogen, oxygen, and nitrogen, to which there must almost always be added—sulphur, phosphorus, chlorine, sodium, potassium, calcium, magnesium, and iron. Thus it is a question not of the presence of rare elements but of the combinations of common ones. The chemist also tells us that the dead protoplasm shows a mixture of proteins, carbohydrates, and fats besides minute representation of some other materials. It is well known that proteins are never absent, and it follows that they must play an essential part in the chemistry of life. Proteins contain carbon, oxygen, nitrogen, and hydrogen (in that order of percentage), and frequently a minute representation of sulphur. They are familiarly illustrated by the albumen of white of egg, the vitellin of yolk of egg, the casein of cheese, and the gluten of the grains of wheat. The protein molecules are very large and complex, sometimes consisting of thousands of atoms, and there are so many different kinds that it is safe to say that every distinctive type of animal or plant has a peculiar protein, characteristic of itself. The red colouring matter of the blood—the hæmoglobin—is a protein, and we know that this pigment in a dog is appreciably different from that which occurs in a horse or a man. Some day, perhaps, when science is nearer defining what is meant by a particular kind or species, the description will begin with a formula of the particular kind of protein, and end with an appreciation of the creature's psychical mood, if it has got the length of having one.

But to return to protoplasm, "the physical basis of life": it contains innumerable invisible particles of protein and other organic compounds in the liquid medium, and there may be crowds of immiscible droplets as well; and all this multitudinousness of minutiæ means a very large development of surface in proportion to the total mass. This allows of a great intensity of changes, because the area of surface is so enormous. Thus there is usually an electric charge on the contact-surface between any two "phases," e.g. between a complex solid particle and a complex liquid medium; and the huge number of the sometimes quivering particles —quivering because bombarded by the restless molecules of the fluid—means a very large surface and therefore a copious spring of electrification. It seems that many of the marvellous properties of living matter, such as rapidity of chemical change, are wrapped up with this "colloidal state," where innumerable ultra-microscopic particles and droplets are suspended in a complex liquid medium.

A cell is a unit-mass or area of living matter usually centred in a kernel or nucleus, but in the laboratory of a cell there is often a production of chemical substances that cannot be regarded as being very intimately connected with the living matter itself. Thus a cell may make granules of a black pigment (melanin) or crystal-like spangles of guanin, and all such things, taken by themselves, must be regarded as side issues, away from the essential chemical routine, which is technically called *metabolism*. Now, if we could subtract from the total of the cell-substance (or cytoplasm) all these by-products, there would be left the genuine living matter. In technical language, cytoplasm minus metaplasm leaves protoplasm.

This is theoretically clear, but practically impossible. The probability is that there is no one substance which should be called protoplasm, but rather a combination of proteins, carbohydrates, fats, and other stuffs which work into one another's hands in a very effective way. There is much to be said for using the metaphor of a firm; the partners are effective in themselves, but the characteristic efficiency of the firm is due to their correlation, that is to say to their power of working into one another's hands. But while we know that a common purpose is the bond that keeps the firm together, able to act as a very successful unity, we do not really know what the bond of union is in a cell. The persistent self-preservative activity of protoplasm is much more difficult to understand than the suicidal explosion of the gunpowder. In the protoplasm there is for a time a succession of up-building, constructive, or *anabolic* processes. But the clock that is thus wound up is ready to run down; and thus there comes to be a succession of disruptive, destructive, or *katabolic* processes. These two sets of processes, winding-up and running-down, make up the metabolism of protoplasm; and it is characteristic of life that the see-saw can be kept up for days or for years or for stretching cycles of years. The living organism is a going concern; its accounts balance.

Perhaps the greatest change of recent years in our picture of protoplasm is that implied in the abandonment of the view that it has a microscopically demonstrable structure, network-like, or like a tangle of fibrils, or otherwise. The appearance of intricate structure has often been seen under high magnification in fixed and stained cells, and many beautiful drawings of the protoplasmic network have been published. But this microscopically demonstrable structure is an artificial product; it does not exist in living cells; it can be mimicked in white of egg. Protoplasm is a liquid emulsion, mostly consisting of water; yet it can be firm as a jelly-fish is firm, and it can make for itself a supporting framework. But when we speak of protoplasm as structureless, we must hasten to add that the cell may be divided into minute areas by very delicate filmy partitions. These extremely delicate partitioning films have diffusion-hindering properties, which allow dissimilar chemical processes to occur cheek by jowl. It must be admitted however that the partitions are not usually demonstrable in any direct way. The old picture was a visible intricacy, which was an "artifact," that is to say, an artificially produced after-death effect; we have exchanged this for an intricacy that is usually invisible, yet indubitably real. How can we be sure of the presence of innumerable invisible particles in a colloid like protoplasm? The answer may be found by going back again to the dust-motes moving in the air of a shaded room through which a beam of light passes. The motes are too small to be seen, but we see the haloes of light reflected from their surfaces. So by means of the "ultra-microscope" we can see the haloes of particles which are themselves too small to be visible. In this way the presence of the microbes causing certain diseases has been *proved* though these microbes are so much smaller than ordinary bacteria that they cannot be seen under the highest magnification.

Let us try to get just a glimpse of the intricacy of life. The simplest living creatures, whether plants or animals, are single cells. That is to say, they are minute units of living matter, each usually centred in a kernel or nucleus. A

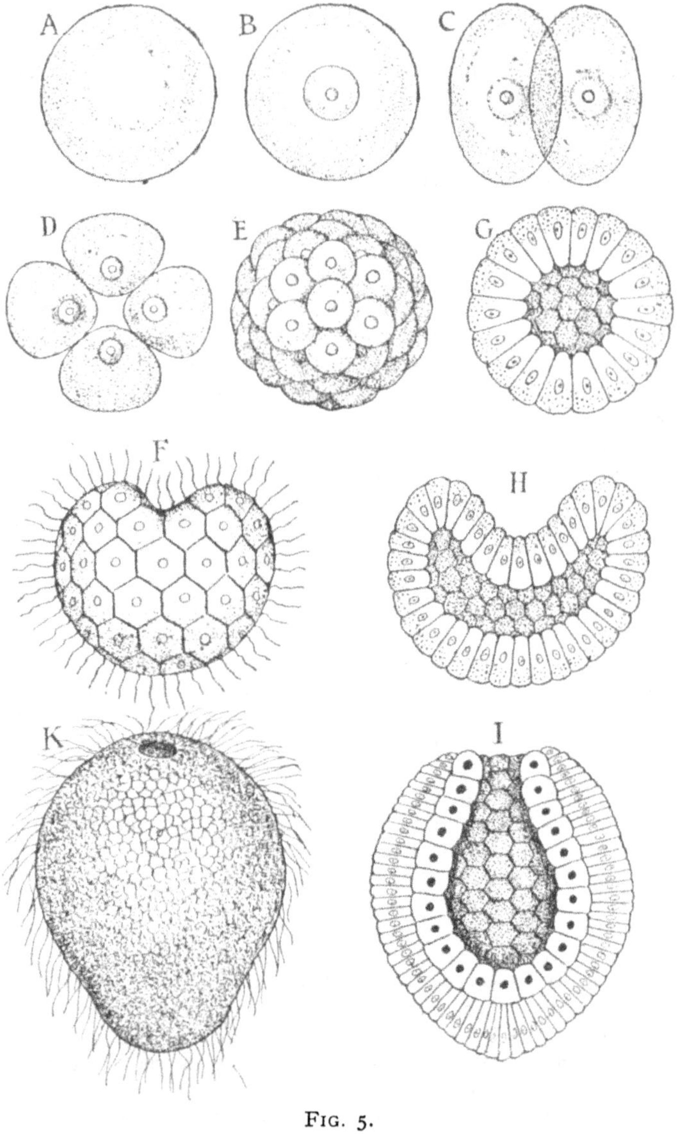

FIG. 5.

Early stages in the development of a coral. After Haeckel. The fertilised egg-cell (B) divides into two (C), into four (D), into a ball of cells or blastula (E), seen in section at G. The ball of cells (E) becomes indimpled or invaginated to form a sac of cells or gastrula (F), seen in section at H. The free-swimming ciliated gastrula is shown at K, and in section at I, where two embryonic layers of cells—ectoderm and endoderm—enclose a primitive gut-cavity.

common amœba from a ditch may be a hundredth of an inch in diameter, and it is a relatively large cell.

All but the simplest living creatures are multicellular, being built up of many cells, each a minute corpuscle or area of living matter, centred in a nucleus. One of the very small many-celled Wheel-Animalcules called Hydatina, that can swim through the eye of a medium-sized needle, has about a thousand cells in its tiny body ; but we have many millions. In all living creatures that multiply in the usual way the individual life begins in a single cell,—the fertilised egg-cell. This divides and re-divides to build up a ball of cells, and so the development of the embryo proceeds. But besides the multiplication of cells there is a puzzling process called differentiation in the course of which the cells become different from one another. In other words, division of labour sets in. Some cells become nervous, others muscular, others glandular, others skeletal, others connective, and so on. If we compare a living body to a city, then a tissue, like muscular or nervous tissue, would correspond to a street of similar shops, each with the same kind of activities.

Cells have been studied for nearly a hundred years, but it is only in this generation that biologists have realised what complex little worlds they are. There is, first of all, the minute corpuscle or area of living matter or protoplasm, which is in a colloidal state. It is a subtle mixture of proteins, carbohydrates, fats, ferments, and so forth, surrounded by a delicate membrane, which in plants becomes a substantial cell-wall of cellulose. Much depends on the fact that this cell-membrane is permeable, inwards and outwards, to certain substances and not to others ; and that this permeability changes with the internal and external conditions. In many cases the colloidal cell-substance includes not only the protoplasm—which Huxley called " the physical basis of life "—but non-living inclusions such as crystals or pigment-granules or globules of oil, and so forth.

In the centre of what is really a rather turbulent whirlpool of complex chemical substances, there floats the nucleus, a microcosm in itself. Inside its membrane, through which materials are always seeping in and out, there are visible, in an active cell, several readily stainable bodies called chromosomes, which are usually definite in number for each species. The smallest number is two, which occurs in the cells of the common threadworm parasite of the horse, whereas the number for the horse itself is 64. Man's number is 48, but there is not much in the number, as is plain when we mention that 48 is also characteristic of some snails and some plantains! The important fact is that the specific number, whatever it may be, is adhered to in all the cells of the body with the single exception of the ripe egg-cells and sperm-cells which have *half* the normal number. When the egg-cell is fertilised by a sperm-cell, the normal number is restored. Another interesting point is that related species of plants and of animals sometimes have numbers of chromosomes which occur in a regular arithmetical series. Thus one species of rose may have 7, another 14, another 28, and

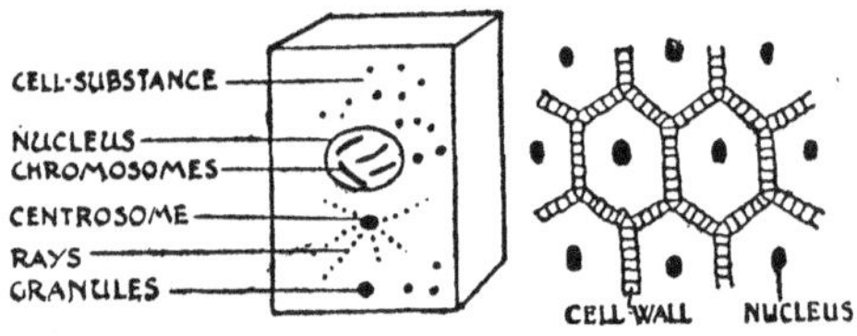

FIG. 6.

To the left a diagram of a cell, seen in its three dimensions. To the right a surface view of a group of plant cells, showing bridges of living matter passing through the walls and connecting cell with cell.

so on. The chromosomes look very solid when they are fixed and stained, but in life they seem to be more like very fluid sausages inside a film. Under high magnification it sometimes seems as if they were built up of a double row of beads, which have been called microsomes. But the great interest of the chromosomes is that they are the bearers or vehicles of the hereditary characters, or of some of them at least. When a cell divides into two cells, as happens in all development and growth, each chromosome is meticulously split longitudinally, so that each daughter-cell gets a precise half of what each chromosome contained.

We have not nearly completed the inventory of what the typical cell contains, but that is beyond our purpose of simply giving a glimpse of the intricacy of life. But we must not end without recalling that besides intricacy of structure there is intricacy of activity. Each cell is a living laboratory in which there take place oxidations and reductions, hydrations and dehydrations, fermentations and condensations, often with extraordinary speed and with not less characteristic orderliness. The cell is fearfully and wonderfully made !

CHAPTER XI

THE PEOPLING OF THE SEA

AS long as we study changes of matter and energy that have nothing directly to do with life, we are studying the domain of things, for which the word COSMOSPHERE has been proposed. In other words, we are studying such sciences as Astronomy, Geology, Chemistry, and Physics. But in our last study we made the step from the domain of things into the realm of organisms—it seems of value to clear thinking to use different words like "domain" and "realm." We passed from the not-living to the living, and for the realm of living creatures the word BIOSPHERE has been proposed. Several studies must now be devoted to the spreading of life over sea and land.

It was perhaps in a quiet sunlit pool that living matter was first built up, and that living specks united to form minute ORGANISMS—the first living creatures. It is well known that gathering together has often occurred in the course of the history of the world; electrons and protons gathered together long ago to form atoms; atoms combined to make molecules; molecules became associated into the micellæ (particles and droplets) that occur, suspended in a fluid medium, in what is called the colloidal state of matter. There may have been another momentous gathering together when particles of living matter cohered to make a little unit or cell. But we have already confessed that we do not *know* anything about the origin of the first living creatures.

It may have been, we say, in a genial sheltered pool that they began, but it is very likely that the first home of living creatures was in the OPEN SEA. Why is this likely? In the first place, it is believed by some investigators that there were expanses of deep open water before there were shallows and shores. In the second place, the Open Sea is a safer cradle than the rough-and-tumble shallow waters where the waves break. The conditions of life are more uniform in the Open Sea than in any other haunt of life, except the Deep Sea. The differences between day and night, between summer and winter, are less marked in the Open Sea than anywhere else. There is abundant sunshine without any risk of drought, and storms can be readily avoided if the animals sink down for a few fathoms. There is room for all in the Open Sea and nothing to knock against except the coral-covered shoulders of an oceanic volcano or the cold sides of a drifting iceberg It is interesting to notice that an immense number of delicate young animals live in the expanse of the open waters, although their later life is near the boisterous shore. This suggests that the

FIG. 7.

The circulation of matter illustrated. Not drawn to scale. 1. Peridinid Infusorians of the Open Sea. 2. Pelagic Copepods, which feed on the Infusorians and also on Diatoms. 3. Mackerel, which feed largely on the Copepods. The fishermen bring about another embodiment.

Open Sea was the primeval cradle of life. It should be noted that by the Open Sea naturalists always mean the open waters as far down as the light penetrates freely—a few fathoms only, perhaps only a score.

All sorts of animals are now represented in the Open Sea, forming what is known as the PELAGIC FAUNA, but many of them are not what can be called "natives"; they have come to the open waters from elsewhere. That must be true, for instance, of whales, which are descendants of terrestrial mammals, and of turtles which are migrants from land and fresh water. Sea-snakes are sometimes found a hundred miles from land, but they are lung-breathers and could not be thought of for a moment as "native" to the sea. On the other hand, there are very numerous single-celled animals which may have lived in the Open Sea from the first; and there are many simple stinging animals and worms that are characteristic of the open waters and may be regarded as aborigines. But one of the strongest reasons for ranking the Open Sea as the first great haunt of life is the abundance of very simple drifting plants—greenish Algæ—which form what Sir John Murray used to call "the floating sea-meadows." They are more abundant in the open waters than anywhere else, and their importance is that they build up air, water, and salts into carbon-compounds on which animals can feed.

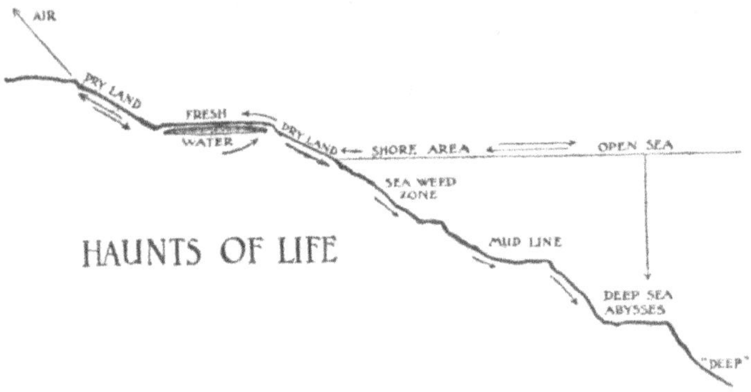

FIG. 8.

The arrows indicate how living creatures have passed from one haunt of life to another, e.g. from shore to deep waters, from shore to open sea, from shore to dry land.

The myriads of animals in the Open Sea—the drifting "Plankton," like jelly-fishes, and the swimming "Nekton," like mackerel—are dependent on the inexhaustible numbers of minute Algæ, which form the fundamental food-supply for higher forms of life.

It is highly probable that the first peopling of the Open Sea was by minute living creatures called Protists, which are, as it were, undecided whether to become plants or animals, yet have the plant's characteristic power of using the energy of the sunlight to build up carbon-compounds, such as sugar, from the air, water, and salts of the surrounding medium. There are many such creatures still very abundant both in fresh and salt waters. Such, for instance, are the Peridinids and the green Flagellates, which are sometimes claimed by the botanists and sometimes by the zoologists. In all probability the original colonists of the Open Sea were without the complex green pigments (chlorophylls) that are found in plants to-day, but had something rather simpler, though with the same property of utilising some of the rays of sunlight in chemical construction or synthesis.

THE SHALLOW SEA

Many naturalists believe that the original home of life was the shore-area—the shallow well-lighted waters where seaweeds grow abundantly. For this view there is much to be said, though perhaps the case is not so strong as for the Open Sea. There is no doubt that the shore is a very stimulating place, with a great variety of changeful conditions,—just the kind of place where life might send out its first tendrils. It is the meeting-place of air, water, and land. Ebb and flow of tides, fresh-water floods at one time and at another the scorching heat of the sun, the alternation of day and night, summer and winter, felt much more markedly than on the Open Sea, the endless variations between gently lapping waves and blasting breakers, the slow sinking and raising of the coast or the sea, these are some of the vicissitudes to which the creatures of the shore are exposed. It must be noticed that naturalists mean by the LITTORAL FAUNA much more than the animals to be found between tide-marks, for the seashore includes the whole of the relatively shallow water around the land, as far as the sunlight penetrates readily and encourages the growth of seaweeds. The strict littoral zone with its limpets and periwinkles, green sea-lettuce and brown bladder-wrack, goes down as far as the lowest tide-mark. Next comes the Laminarian zone, named after the great pennon-like brown seaweeds, which wave over a fierce battlefield of starfishes and sea-slugs, crabs and sea-urchins. Farther down the slope, from 15-40 fathoms, is the Coralline zone, where the seaweeds are mostly red in colour or very stiff with lime, where giant whelks prowl about in search of booty and are themselves devoured by rock-cod and other inshore fishes.

There can be no doubt that the shore-area has been a great school of life, where most of the main groups of animals have learned invaluable lessons such as alert seizure of opportunities and patient endurance until the tide turns. It is also certain that all the chief groups of animals are now represented in the seashore-area. The strongest reason for not thinking of the shore as the original cradle of life is, as we have said, the difficulty of the situation; there is so much change and turbulence. Living creatures have extraordinary persistence, endurance, and even insurgence, yet, to begin with, they probably had very delicate constitutions, for which the Open Sea is better suited than the stimulating shore.

The Deep Sea

Long ages passed before animals did much in the way of colonising the Deep Sea, by which is meant the floor of the ocean and the dark waters in the abysses which are often miles below the surface. The probability is that there were not many deep-sea animals before the Cretaceous Ages, and that even then the peopling of the greater depths was very gradual. We cannot wonder at this when we think of the strangeness of the Deep Sea. Apart from the fitful gleams of "phosphorescent" animals, it is a world of darkness, of eternal night, for even sensitive photographic plates are not influenced when automatically exposed below 250-500 fathoms. It is a remote world, though it is easy enough to throw a stone into it. The average depth of the ocean is 2½ miles, and there are great "deeps," as they are called, of over six miles, in which Mount Everest might be hidden away! There must consequently be enormous pressure in these ocean depths, 2½ tons per square inch at 2,500 fathoms, a depth at which wood sunk down by leaden weights is so compressed that when it is pulled up again, it will not float. By this weight of ocean the abyssal animals would all be squeezed out of existence were they not so interpenetrated by water that the great pressure is not felt, being the same inside as outside.

The Deep Sea is extremely cold, about the freezing-point of fresh water, an eternal winter. The sun's heat is virtually lost at about 150 fathoms, and there is from the polar regions a continual sinking down of cold water rich in oxygen. It is a very monotonous and uniform haunt of life—eternal winter, eternal night, and a lasting calm, for there are no rapid currents, and even the severest storms are relatively shallow in their reach. There are no plants except perhaps the resting stages of some Algæ, for typical vegetation depends on light. Even bacteria, otherwise almost omnipresent throughout the earth and its waters, do not seem to flourish in the great depths. Thus there is no rotting in the Deep Sea, and the carcase of a whale, compressed like pemmican, is simply nibbled away by crustaceans till only the stone-like ear-bones are left. And if there are no plants, the abyssal animals must feed on one another, or must depend on the slow rain of minute organisms and organic particles from the surface waters overhead.

It is likely that the ABYSSAL FAUNA is mostly made up of migrants from the shore-area, which followed the drift of food down the long slope. Perhaps there was a particularly abundant colonisation from the two Polar Regions; and it is also probable that some other colonists sank down from the open sea into the ascetic and unstimulating retreat afforded by the deep waters.

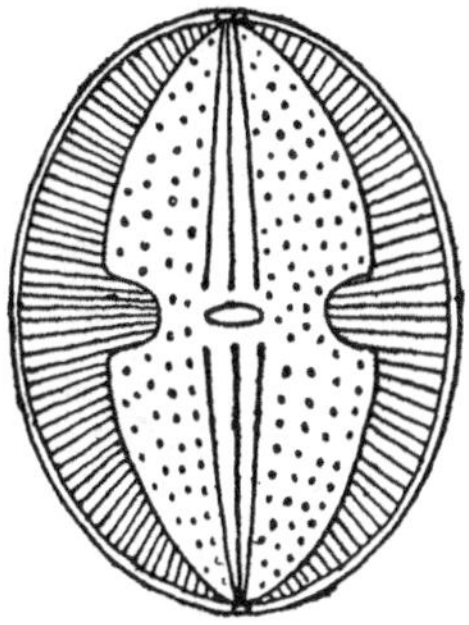

FIG. 9.

A single Diatom, a microscopic single-celled plant with a shell of flint, which is beautifully sculptured with lines and dots.

There are striking contrasts between these three marine haunts of life—the Open Sea, the Shore, and the Abysses. The shore waters are characteristically changeful, the abysses are monotonous, the open sea is between the two. The shore-area is much more crowded than the two others, and ever so much smaller in extent, though of enormous length.

As regards food-supply, the life of the Open Sea depends in the long run, as we have said, on the floating population of simple plants, but the nutritive chains are sometimes long. Thus the dolphin may devour the mackerel, which feeds largely on minute Copepod crustaceans, and they again depend on microscopic animals, such as Infusorians, and microscopic plants, such as Diatoms. Thus the supply of mackerel on the market varies with the sunshine record, for the sunlight favours the multiplication of the unicellular plants in the sea.

Deep-sea fishes may feed on deep-sea crustaceans, which devour deep-sea worms, which depend on the rain of minute organisms from the surface waters overhead. As there are no plants in the abysses, the often multitudinous life would be unintelligible were there not that outside source of food, sinking down from above. Every kind of dead creature is devoured that sinks into the abysses, except

FIG. 10.—MACKEREL, *SCOMBER SCOMBRUS*.

the quite inedible parts, such as the ear-bones of whales and the teeth of sharks.

In the shallow water or shore-area, there are often short nutritive chains. Thus the periwinkle browses on the seaweed, and so do the inshore turtles, in contrast to those of the open sea, which are fish-eaters. But the food chain is often long; thus the octopus devours the crab, which feeds on the worm, which depends on the fine debris and on microscopic organisms. It is interesting to follow some of the common transformations of matter. A pound of cod-fish means that the cod must have eaten ten pounds of whelk; a pound of whelk means ten pounds of worms;

and a pound of worms means ten pounds of minutiæ, which we may call sea-dust. Thus if a hungry man eats a pound of cod to dinner, he is eating a thousand pounds of sea-dust. So the world goes round, from one incarnation to another. This is the *circulation of matter*.

It is useful to select a few of the fitnesses that mark the animals of each haunt. Thus pelagic or open-sea animals are usually lightly built, very watery, with great power of flotation, and often translucent. Deep-sea animals are often raised out of the ooze on stalks, or move about delicately on long stilt-like limbs. Their bodies are readily interpenetrated by water, so that the enormous pressure is not felt. The sense of touch is in many cases exquisitely developed, as one would expect in a world of darkness. Shore animals show much in the way of armour and weapons. Many are fixed, many take a grip of rocks and seaweeds, while those that move quickly have usually some power of taking hold rapidly—so that they are not carried out to sea by the ebbing tide. Not a few, like starfishes, brittle-stars, and crabs are able to save their life by surrendering a part, which can be regrown at leisure. This self-mutilation or autotomy is not only life-saving; it may be utilised when the animal has been badly wounded in the fray or when it is seriously damaged by the accidental maiming that is apt to occur when stones are shifted about by the waves. It will be a good exercise to multiply these instances of fitness or "adaptation" by five, for every haunt of life is full of them.

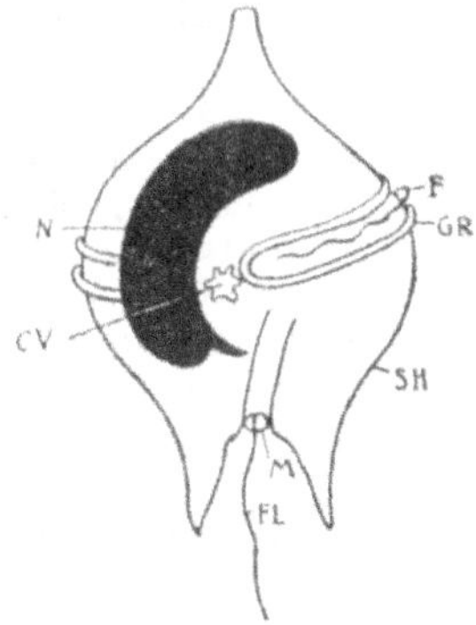

FIG. 11.

A much-enlarged microscopic Peridinid Infusorian, Peridinium (after Schütt). N, the nucleus; CV, contractile vacuole; F, an undulating flagellum, moving in a transverse groove (GR); SH, the shell composed of firm plates of cellulose; M, the end of a longitudinal groove from which there projects a locomotor flagellum (FL).

CHAPTER XII

THE FRESH WATERS AND THE DRY LAND

THERE are six great haunts of life—the shore of the sea, the open sea, and the depths of the ocean, the fresh waters, the dry land, and the air with its birds and bats and butterflies. We have studied a little the three haunts of the sea; and now we must follow the animals, and the plants too, from salt water to fresh, and then on to the dry land. It must be understood that there was not a single great emigration out of the sea—things rarely happen that way; there were many emigrations at different times and undertaken by different races.

There is a very interesting population of plants and animals in lakes and rivers, ponds and pools, bogs and swamps, and other haunts which may be included under the wide title of "fresh waters." Some of them, such as fresh-water sponges and polyps, are doubtless derived from marine ancestors. Some may have worked their way gradually from the shore to brackish water, and thence into river and lake. Some small forms, able to adjust their constitution to new conditions of life, may have been able to survive a sudden change from salt water to fresh, and may have been carried on the feet of birds or on the wings of the wind. A third set of fresh-water organisms may owe their origin to the conversion of a land-locked sea into a lake, or of a blocked-up estuary into a fresh-water loch. When the salt water became fresh a certain number of organisms, sticklebacks for instance, might be able to survive the change. In the fourth place, we must take account of energetic animals which have *actively* colonised the rivers and lakes, from the sea. Thus flounders may flourish in rivers many miles inland, though they must descend to the sea to spawn and to develop; and the Common Eel is probably a deep-water marine fish which has found it profitable to explore the fresh waters. But there is yet another and quite different origin of fresh-water animals, namely from the dry land, as in the case of terrapins and water-beetles, or even from the air, as in the case of many insects like mosquitoes, dragon-flies, and may-flies, whose larval life has come to be aquatic, though the aerial adults are singularly emancipated—even from the dry land.

The Dry Land

Life is much more difficult on the dry land than in the waters, but animals have used the difficulties as opportunities for progress. Animals in the water have great freedom of movement, up or down, to right or left, forwards or backwards, but on the surface of the earth, they can only move horizontally, unless they become burrowers or jumpers. Thus the movements on land have to be very quick and precise when danger threatens, and this favours the evolution of rapidly acting muscles and alert brains for the control of these. It is easier for a fish to escape in the water than for a fox to escape on a field; better three dimensions than two. In water, again, there is no risk of the animal being scorched or dried up, and the moist surface of the body or of the gills is well suited for the in-passage of oxygen and the out-passage of carbonic acid gas. On land, however, it is usually important that the skin should be protected unless the animal keeps to moist and shaded places, or lies in shelter till the sun goes down. But the thickening of the skin or its protection under some extra covering implies a lessening of its power of gaseous interchange. This makes it easier to understand the evolution of *internal* breathing surfaces, such as the air-tubes of insects or the lungs of reptiles, birds, and mammals. The naked frogs are still able to breathe through their skin in the old-fashioned way, and this is what they do in their winter rest; but they also develop lungs after their gill-breathing tadpole period is over. Betwixt-and-between animals, they and other Amphibians evidently are, with one foot in water and the other on land.

Leaving the water raised a new difficulty, namely the safe disposal of the eggs. For while the water affords a universal cradle for delicate germs of life, the dry land involves exposure to drought and the risk of being trampled on or devoured. Thus we can understand why land animals should discover ways and means of securing the safety of their eggs, hiding them in deep burrows, or laying them in lofty nests, or carrying them about in bags as many spiders do, or taking them back to the water again, as in the case of frogs and toads. The best solution of all is to allow the eggs to develop within the body of the mother (*viviparous birth*), for in this way it comes about that what is launched on the voyage of life is already well equipped. We see then that when marine or fresh-water animals ventured to explore the dry land, they had to face many difficulties. When these were conquered, which did not always happen, a door was opened to a more masterful kind of life.

The first colonists of the dry land were worm-like animals, and the most noteworthy achievement was on the part of the earthworms which discovered the possibilities of the under-world. In the course of ages their industry on their own behalf led to what was of incalculable importance to other creatures—both plants and animals—*the making of fertile soil.*

The second great invasion was on the part of simple air-breathing jointed-footed animals (Arthropods), hints of which we get in an old-fashioned survivor called Peripatus, which is widely distributed in many parts of the world. It is in some respects reminiscent of ringed worms or Annelids, but it points on to centipedes and millipedes, and eventually to the extraordinarily successful class of insects. One of the great results of this colonisation was the most important linkage in the world, that between flowering plants and the pollen-carrying insect-visitors, which secure cross-fertilisation. It is probable that independent invasions on the part of other jointed-footed or Arthropod animals, led to the land crustaceans such as the robber-crab that climbs coco-palms on coral islands, and the flat wood-lice that run about under bark and among loose stones. Another allied invasion led to spiders, scorpions, and mites. In all cases it is interesting to look out for animals that have *returned* to the water, just as if they found the conditions of life on the dry land too difficult. Thus the fascinating fresh-water spider, though by no means lacking in something like courage, is an instance of a *return* to an aquatic haunt. So it is with many insects, like the water-measurers that skate about on the surface-film, or like the great water-beetles and the indefatigable little whirligigs. Most striking of all are the members of the family of sea-skimmers (Halobatidæ), insects that have found a home in the open ocean! There are also a great many water-mites, but it is just possible that they are clinging to the ancient home of their race, and are not the result of a retreat from land to water.

Passing over some minor invasions of the dry land, such

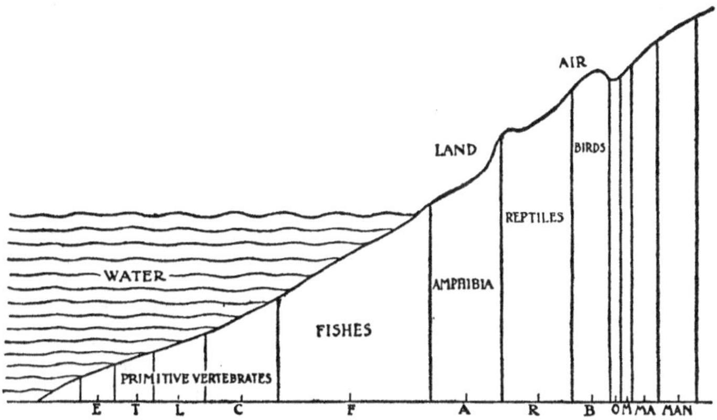

FIG. 12.—DIAGRAM OF THE ASCENT OF VERTEBRATES.

The perpendiculars indicate the progress of animal life,—in differentiation and integration. E, Enteropneusts, worm-like forms with some Vertebrate affinities; T, Tunicates or Sea-squirts, near the base of the Vertebrate series; L, Lancelets, primitive pioneer Vertebrates; C, Cyclostomes, jawless, limbless, scaleless old-fashioned types; F, Fishes; A, Amphibians; R, Reptiles; B, Birds; O, Oviparous mammals; M, Marsupials; Ma, ordinary Placental Mammals; *MAN*.

as that which led to our all too familiar terrestrial snails and slugs, we may notice as the third great conquest that achieved by the Amphibians. Arising from a fish ancestry, the amphibians were the first backboned animals to get a footing on terra firma. Some of them, like the newts, remain in great part aquatic, though normally lung-breathers; and all but a few tree-toads and the like have to return to the water at the breeding season. For a couple of months or more, tadpoles breathe by gills, after the fashion of fishes, though it must be noted that their gills are more like the external gills of the mud-fishes than those of the ordinary fish type. Part of the interest of watching the development of pollywogs in spring is that they show the frog climbing up its own genealogical tree. They are recapitulating in the course of a few weeks the epoch-making advance that the pioneer amphibians made, many million years ago, in passing from water to land. We use the high-sounding word " epoch-making " because the conquest of the dry land by amphibians led on to the evolution of Reptiles, from which in later ages both Birds and Mammals took their origin.

Aerial Animals

The last region to be conquered was the air, and although no living creature is wholly aerial, there are some that have attained to practical independence of the earth. Picture the swift in its headlong flight, never touching earth, save at the nest, from dawn to dusk of the long summer day. Some of the insects, like the Mayflies, which spend their early life in the water, never touch earth during their short adult ecstasy, which may not last for more than one evening.

The difficulties of life on dry land have been referred to, and it is easy to think of circumstances, such as floods, lava-flows, sand-storms, and long-continued drought, which would make it very desirable to get off the surface of the earth. In many cases there may have been ages of climbing and jumping, flopping and swooping, before there was true flight, which implies striking the air with wings of some sort. So to-day the flying-fishes take great leaps above the foam, holding their enlarged pectoral fins taut or vibrating them very rapidly without real strokes. Swooping from branch to branch is illustrated by the web-footed tree-frogs and by various lizards, such as the little flying Dragon (Draco volans) of the Far East, which has its skin, between fore and hind limbs, spread out on five or six much elongated ribs which fold in when the creature is at rest. The same sort of parachuting is seen among mammals, and must have been attained along different lines of evolution, since it occurs in unrelated types like the " Flying Opossum," the " Flying Squirrel " and the " Flying Fox."

True flight has been achieved four times, and on four

distinct lines. In insects the wings are flattened outgrowths, usually two pairs, which are moved with great rapidity, sometimes with 200 beats in a second, by very powerful muscles. In the extinct Flying Dragons or Pterodactyl Reptiles, the wing was a fold of leathery skin drawn out on the greatly elongated outermost finger. In birds, the wing is of course mainly due to the feathers supported on the transformed fore-limb. In bats, the wing is a skin-wing, supported mainly by the much elongated fingers, but carried down the sides of the body to the hind-legs, and beyond that to the tail if there is one.

Looking backwards, then, we see that there are six great haunts of life—the Open Sea, the Shore of the Sea, the Deep Sea, the Fresh Waters, the Dry Land, and the Air, with its birds and insects. But there are some *minor haunts* of great interest. Thus there is the under-world beneath the ground—a refuge which was discovered ages ago by the earthworms, who were followed by centipedes, burrowing beetles, burrowing slugs, blind worms, slow-worms, and, long afterwards, by the half-blind moles. Mr. Edmund Blunden had a fine vision of them when he wrote his " Gods of the Earth Beneath " :—

> I am the god of things that burrow and creep,
> Slow-worms and glow-worms, mould-warps working late,
> Emmets and lizards, hollow-haunting toads,
> Adders and effets, ground-wasps ravenous ;
> After his kind the weasel does me homage,
> And even surly badger and brown fox
> Are faithful in a thousand things to me.

Very different from the underground burrowers, which are usually strenuous creatures, are the " troglodytes " that live in caves. For many of them are infirm, many are weak-eyed, many are nervous and delicate. A good example is the wan Proteus, a blind newt that lives in the underground waters of the Dalmatian caves. It has a near relative, Typhlomolge, as far away as North America. Similarly there are a few cave-fishes, cave-crayfish, cave-beetles, which are never seen in the open world. Perhaps some of them have become weakly because they have lived so long in caves, but there is much to be said for the view that most of the cave animals took to this refuge because they were weakly. The Old Testament cave of Adullam was a retreat for strong, discontented, desperate men, but that is not the character of the cave animals, which are usually handicapped by some weakness or infirmity. Along with the caves might be included mines, which have come to be the haunts of more animals than one would suspect. But the tenants of the mines are newcomers compared with the troglodytes who have lived in caves for ages.

Another home of life, well worthy of being considered by itself, is in the trees. It is a fine study to take an isolated tree in the middle of a field, and make a census of all the animals that one finds there throughout the year, not as casual visitors, but as tenants for a considerable time. There may be a squirrel, a tree-creeper, possibly a tree-snake, possibly a tree-frog, certainly some slug or snail, many a spider and mite, scores of insects, the wood-lice below the bark, and even an earthworm or two in the mould that gathers in a big cleft. Some of the animals that have tried arboreal life are very remarkable. Thus in warm countries the little land-leeches sometimes drop from the branches; the skip-jack fish, Periophthalmus, often creeps on the knee-like above-ground roots of the mangrove-trees; the robber-crab climbs up the coco-palms that have grown on Christmas Island. There is no end to the *insurgence* of life—its tireless search for new homes.

Getting on to the trees seems to have been of great importance in the history of the highest mammals, for there is a long ascending series from the Tree-Shrew (Tupaia) to the Spectral Tarsier (Tarsius), from Lemurs to Marmosets, from New World monkeys to baboons, and thence to apes. What the series tells us is this, that when the climbing mammals got *a free hand*, no longer merely a supporting fore-leg, when they began to grasp and manipulate things, touch and vision came to count for more, and the sense of smell for less, and there was a great improvement in the brain. Great consequences followed the long arboreal apprenticeship—it led the way to Man; though the most critical time was probably when Man's ancestors left the trees and tried their fortunes on the ground.

CHAPTER XIII

THE MARCH OF THE SEASONS

SPRING AND SUMMER

IN the last two studies (XI and XII), we have illustrated one of the large facts of life—the way in which living creatures tend to fill up every corner of earth and sea. Thus there are many plants and animals in the barren grounds or Arctic tundra, snow-covered and frost-bound for eight or nine months of the year ; there is a very distinct fauna and flora of the high Alps, where snow voles and blue gentians are at home ; there is a dense population of animals in the unbroken night and endless winter of the Deep Sea.

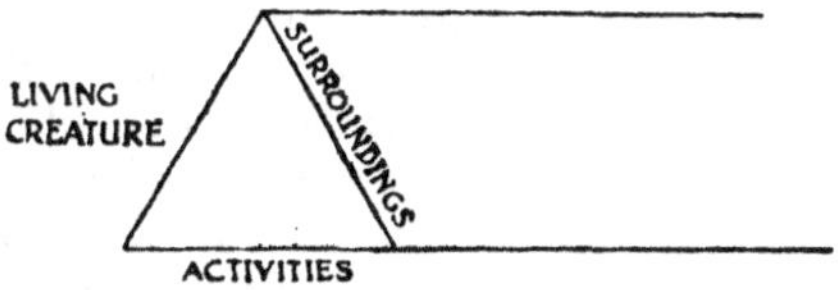

FIG. 13.—THE BIOLOGICAL PRISM.

The organism must always be studied in its relations to its surroundings or environment. It acts on its surroundings and is acted on by them. The prism is drawn uncompleted, to suggest that all three sides are changing. It is not the same organism, or the same environment, or the same functioning, all the time.

But another great fact, the study of which is often called PHENOLOGY, is the waxing and waning of life as the earth moves round the sun. In other words, there is a march of the seasons, especially well-marked in temperate countries ; and the seasons do to some extent hold living creatures in their grasp.

As the sun is the source of almost all the energies of the earth, our income—of heat and light in particular—varies with our seasonal position in relation to the centre of our system. With this chief relation of temperature and illumination there are associated many minor influences that change with the seasons, such as the rainfall and the winds. In general and in detail, the waxing and waning in the physical world sway the changeful aspects of life in plants and animals. One of the basal facts is that the ratio of heat-supply in summer to that in winter is as 63 : 37.

Spring

Let us begin with spring, the season of young things and re-awakening, of renewed activity and vigour. The water in the soil begins to move again, for the spring sunshine draws up the mist. There cannot be any life, such as we know, where there is not water in a liquid form, and the earth is the only planet about which this can be said with certainty. More than once we have noted that living matter or protoplasm always contains at least 70 per cent. of water; and it is therefore of fundamental importance that water should keep circulating. The continuance of life depends on what is learnedly called "the meteorological cycle," that is to say the circulation of water in its various forms. The mist rises from the deep, the water-vapour is drawn up in the warmed air; clouds are formed and shepherded across the sky by the wind, or more accurately there are currents of air in which, according to the temperature, clouds appear and disappear and appear again. Against the cold shoulders of the mountains the air-currents are cooled, and the water-vapour becomes droplets of water on the rocks. The droplets form drops and the drops unite into runlets, which are often delayed for a time by the thirsty lichens high up on the hills beyond even the heather zone. But the runlets from condensed water-vapour are helped by rain and snow from the clouds, and so the streamlets are fed, often so abundantly that they form little torrents which hurry down the mountain side, making long scars. Anything that lessens this over-hurry of water is useful, and we should always remember with gratitude the bog-mosses (Sphagnums) that form great sponges, acre after acre, at the foot of the hills. It is mainly because of these great sponges that the springs continue welling and the streams continue flowing in dry seasons, even when there has not been rain for weeks. But we need not continue further along this

PLATE II

LAMBS AT PLAY

line of thought. Much of the water is captured by plants and animals, some being delayed in their bodies for a long time, some being split up altogether (so that the hydrogen and the oxygen of the water find new partners) ; and some being soon given off again as water-vapour into the air. But sooner or later much of the water finds its way back to the sea, often carrying salts filched from the rocks and the ground, and thus the "meteorological cycle" is completed. Let us notice then that one of the fundamental facts in regard to spring is the increasing warmth of the sunshine, and the consequent quickening of the cycle of mist and cloud, rain, river, and sea.

Formed in the warmth of last summer's sunshine, scattered and sown in diverse ways in autumn, myriads of seeds have been resting in the soil throughout the winter. Processes of fermentation have been going on for part of the time, changing the very condensed store of food into more usable form; thus proteins become peptones and starch becomes sugar. After some months, it may be, the envelopes of the seeds begin to crack, and the microbes of the soil find their way into crevices, so that rotting begins. "Except a seed of corn fall into the ground and die, it abideth alone; but if it die it bringeth forth much fruit." The process of dying here referred to, which liberates the seed-life, is the decay of the seed-coats. Now one of the great events of spring is that the water in the soil seeps into the seeds whose envelopes are being opened; for this water helps in the fermenting and rotting; it causes swelling; and, most important of all, it enables the living matter or protoplasm of the seed to become active. That is to say, development begins. The embryo in the seed is developed into a seedling, which elbows its way out of the seed-envelopes. The seeds sprout, the stem-part growing upwards and showing its head out of the ground, the root-part growing downwards and taking a grip of the earth, from which it begins to absorb more water. How soon the whole life of mankind would come to an end if the seedlings did not lift their heads above the ground and make the brown earth green in the springtime.

In all the trees there are delicate pipes, as slender as hairs, and therefore known as capillary tubes, which extend from the roots, up the stem, to the leaves. They are called wood-cells (tracheids) and wood-vessels, and those of them that are living and form the young wood or sap-wood are of essential importance because they are the plant's water-pipes. There is a continuous series of them from the roots, where water is absorbed from the soil, to the leaves, where water is used up in making sugar and the like, and also lost into the air in what is called transpiration. From level to level there are relays of these delicate microscopic pipes, containing soil-water ; and when some of the water is used up or is lost in the leaves, more water *must* rise to take its place. If we have a long water-column in a capillary glass tube rising out of a tumbler, and if we remove some of the water at the top, more water is bound to rise to take its place, for there are physical forces of cohesion that make it surprisingly difficult to " break a water-column." Something similar applies to the water-tubes in a plant ; and this is one of the reasons for the ascent of sap, one of the great events of spring.

But we must also notice that if a glass vessel filled with syrup, and closed with parchment paper, be immersed in water, there is a diffusion of the water into the syrup, which will vary in rapidity according to the difference in density between the two. It will go on and on until the syrup is very watery. Now the cells just behind the tips of the rootlets are the chief water-absorbers, and the reason of this is roughly the same as is indicated in the experiment referred to, for the water in the soil must diffuse through the cell-wall into the partly syrupy dense contents of the root-cells. But as these cells have their contents slightly diluted, the cells above them will absorb from them, and so the process continues. This is another reason for the ascent of sap.

It must not be thought that it is an easy thing to account for the ascent of sap, but we have mentioned two of the chief factors. And it must not be supposed that the whole function of the living cells and vessels (a plant-vessel is due to the fusion of a row of cells) in the " green wood " is to secure the up-current of water and salts from the soil, for there seems good reason to believe that some of them serve for the down-current of sugar and proteins from the seat of manufacture in the green leaves. But all this is just to hint at the familiar marvel of the spring, that the water rises against gravity to the tops of the tall trees, with the result that the buds, formed the previous summer, begin to grow and swell and open, scattering their protective bud-scales as litter on the ground. It makes little difference whether the plant is a cedar of Lebanon or a hyssop on the wall, the sap rises. So it comes about that over the farmers' bare fields, over the brown moor, over the dry links, and over the leafless trees of the forest there is stretched in spring a great green veil, which enables the plants in the months that follow to capture part of the power of the sunshine. The continuance of Man's Kingdom depends on this green veil.

Spring is a time of new beginnings, like the seedlings, and of re-awakenings, like the opening buds. The word re-awakening makes us think of the old story of the Sleeping Beauty whom the Prince kissed awake, for the Sleeping Beauty was our fair earth, lying under the spell of winter, and the Prince was the sunshine of spring, which nothing can resist. This naturally leads us to think of the " winter-sleepers " or hibernators among animals, which have lain in a strange state of more or less " suspended animation " through the cold months, and are re-awakened in spring. The only creatures to which the term " hibernation " should be applied are certain mammals, like hedgehog, dormouse, marmot, hamster, and bat. They are what may be called imperfectly warm-blooded, and when it becomes cold in autumn, they cannot produce enough of heat in their bodies to make up for what they are losing into the cool air. So, prompted by inborn instinct, they seek out sheltered nooks, and subside into a strange state of greatly reduced vitality, with no income and very little obvious expenditure. Spring is the time for their re-awakening, often none the worse for their long rest and fast, during which they have lain low and said nothing, as Uncle Remus put it. When we deal with winter, we shall mention other creatures, besides the true hibernators, that sink into a state of suspended animation during the cold months and are re-awakened in spring. We are thinking of animals like the slow-worms, the frogs, the snails, the humble-bees, which pass the winter in a comatose or half-alive state—more or less like the resting seeds. But this kind of " lying low " is quite different from true hibernation.

Just as the winter sleepers return in spring to their old haunts from their hidden retreats, so there is a joyous return of the migrant birds which went South in autumn.

In a North Temperate country the majority of the different kinds of birds are "summer visitors," such as swift and swallow, cuckoo and nightingale, wheatear and warbler. They left us in autumn for winter quarters in the South, often in Africa; they return in spring a merry throng, often with the joy of spring in their voices. The songs, usually confined to the cock-birds, are part of the courtship, for spring is the time of love-making. They court and pair, they nest and brood; and by and by there is a young family requiring much attention. This is another feature of spring, that it is the time of young things, not only the nestling birds, but young mammals (e.g. hares and rabbits), young insects (e.g. grubs and caterpillars), young frogs or tadpoles, and all sorts of animals of high and low degree.

It is interesting to watch a pond when spring comes. First, in the growing sunshine, there is a multiplication of single-celled plants, very simple fresh-water Algæ. There follows a multiplication of likewise microscopic animals, such as some kinds of Infusorians. Then more complicated visible animals begin to be abundant, such as the pinhead-like crustaceans badly called water-fleas. Later on there may be a crowd of much larger animals like water-insects and fishes. Just as there are many young creatures on dry land, so in the pools there appear many "water-babies," as Kingsley called them, such as tadpoles and young gnats. If we try to state the case very generally, we may say that the circulation of matter from one embodiment to another is slowed down in winter, and is quickened again in spring. Grass begins again to become flesh; diatoms begin again to become fishes; one re-incarnation follows another all the world over.

Spring is the time of new beginnings, when many forms of life make a fresh start; the time of re-awakenings from winter's rest; the time when the migrant birds return; the time of re-peopling of wood and meadow, lake and sea. Spring days are the days of youth—seedlings, caterpillars, mayflies, tadpoles, leverets, lambkins, nestlings. April is said to mean "opening," and is it not the time when the frost-bound earth opens, when the seeds open, and the buds and the eggs, the bird-orchestra, and much more besides. For we must not forget the spring flowers, characteristically white and yellow,—the snowdrops and crocuses, the wood-anemones and celandine, the sloe-blossom and the golden glory of the broom. The tide of life began to ebb in autumn ; it reached its low-water mark in winter ; in spring it has turned and we watch it flowing—sometimes very gently, sometimes like a flood.

SUMMER

The biggest fact about summer, as far as living creatures are concerned, is the great increase of income. This income consists mainly of heat and light, and of the food from the air and from the soil which is made available by these radiant energies. And if the income is greatly increased, as compared with what was available in the early days of spring, then there is much more opportunity for expenditure. Thus summer is the time of greatest industry or activity.

Let us make this more concrete. The day has become longer and warmer, and the longer light gives green plants more time to utilise the red-orange-yellow rays, whose energy, passing through a screen of green pigment, enables the leaves to build up sugar and other carbon-compounds. The radiant energy of the rays with short wave-lengths is somehow changed into the chemical energy of carbohydrates and proteins ; and though we do not exactly know what takes place in the living laboratory of the leaf, Professor Baly and his collaborators have shown, as we mentioned before, that light shining on a vessel containing water and carbon dioxide may bring about the up-building of formaldehyde CH_2O, the simplest of all the carbohydrates. Longer exposure may bring about the formation of a sugar, which the students in the laboratory have sometimes used to sweeten their tea.

Whatever the details of the manufacture may be, it is certain that every plant is a sugar factory, what might be called a synthetic sugar factory, building up sugar from carbon dioxide and water. But the synthesis does not stop at starch and sugar, it goes on to the even more valuable food-stuffs called proteins. There is no living matter that does not consist in part of these nitrogenous carbon-compounds, such as the gluten of wheat, the legumin of peas, the albumen of white of egg, and the casein of cheese. Proteins are the most important of all chemical substances, and there are thousands of different kinds, for as St. Paul said, " all flesh is not the same flesh."

Sprouting and leafing are characteristic of plants in spring, but one of the features of summer is the *industry* of the foliage. In perfect silence, all through the sunlit hours, there is in every green leaf a great bustle of chemical molecules and a manufacture of carbohydrates, proteins, and other carbon-compounds. These form the materials for the plant's further growth, for the repair of injuries, for the upkeep of active parts, often for filling the fruits with sweet juice, and for giving the seeds a legacy of nutritive reserves. But in many cases, though not of course in *annuals*, much of the food-material that is manufactured is not used in the present at all, but is stored as capital or savings for the next spring. This is the meaning of the substantial root-stocks or rhizomes that are familiar in, say, irises and bracken ; the nutritive tubers, which are just swollen knobs of an underground stem, so much appreciated in potatoes ; the condensed stem-bases or corms well illustrated by the crocus ; and the swollen underground buds or bulbs of the onion and the hyacinth. All are stores of surplus material which will be used next year. This fact makes it easier to understand the frequently sudden outburst of foliage and flowers in spring. The plants are cashing their legacy.

Unlike animals, plants make their own food ; and they make enough not only for themselves, but for the animal kingdom as well. For in the long run, animals depend on plants for their food. This makes it probable that if summer is the height of industry for plants, it will be the same for animals ; and so it is. The " busy bee " that " improves each shining hour " is a symbol of the summer industries of animals. Prompted by appetite and urge, guided by much instinct and a spice of intelligence, how hard many animals work in getting food, in building shelters, and in accumulating stores. There are hunters like the spiders, fishers like the herons and pelicans, miners like the moles, foresters like the beavers ; there are agricultural ants that harvest and there are others that keep domestic animals. Summer is the time for studying animal industries, and the climax is to be seen in the ant-hill, the termitary, the wasp's nest, and the bee-hive.

But we must not forget that in some of the higher animals there is play as well as work ; and play which begins in spring is continued into summer. It is a safety-valve for

SUMMER INDUSTRY

TWO OLD-FASHIONED BEE-HIVES

overflowing energy and high spirits ; it is an irresponsible opportunity for serving an apprenticeship in the business of life—for play is the young form of work ; and it is an opportunity for testing those new departures or variations in behaviour on which racial progress in part depends. Play is of great biological interest and of quite vital importance ; it is familiar in lambs and kids, calves and foals, kittens and puppies, young foxes and young otters ; and it is happily continued among children, though a little apt to be replaced by conventionalised games.

Autumn for fruits and seed-scattering ; winter for retrenchment and rest ; spring for sprouting and foliage ; summer for *flowers*. No doubt there are many beautiful flowers in spring, but they have not the exuberance of Flora's summer pageant. It is interesting, for instance, to notice how the deeper colours become commoner as the summer months pass. There are more blues and purples, and perhaps these deeper pigments are the by-products of a more intense life.

It is in summer that we see most of what must be called the most important linkage in the world, that between flowers and their insect-visitors. "Most important," we say, because many of these visitors, notably the various kinds of bees, that come for nectar and pollen, carry the fertilising golden dust from one blossom to another blossom of the same kind. It is this fertilising pollen that makes the possible seeds or ovules into real seeds that will sprout. The insect, all unconscious of its services, dusts the moist "stigma" of a flower with pollen of the same species. The pollen-grain sends out a long microscopic pollen-tube, which grows down the "style" towards the egg-cell within the "ovule," within the "ovary." A male nucleus in the pollen-tube unites in an intimate and orderly way with the nucleus of the egg-cell, and this fertilisation is *the beginning of a new individual life.* The fertilised egg-cell divides and re-divides to form an embryo or young plant, in other words—a seed that will sprout when it is sown. Just as the fertilised egg-cell of an ordinary mammal (a hair-covered milk-giving quadruped) develops inside the mother into an embryo, which goes on developing until it is ready to be born, so among flowering plants the fertilised egg-cell develops within the parent into an embryo, which stops at a certain stage and waits till it is sown. In both cases there is a close union between the offspring and the mother-organism, so that there is often a long nurture before birth. It is worth thinking over the fact that in the highest plants as the highest animals, there is *viviparous birth.* That it to say, what is separated from the parent is already a young creature more or less advanced. When we look at the flowery meadow in the summer, we should remember that for many millions of years there were no flowers and no seeds. The more old-fashioned flowerless plants, now represented by ferns and horse-tails and the like, are without seeds. It was a great step when plants became viviparous.

Now let us go back for a moment to the insect-visitors which are so characteristic of summer. In many plants, like pine-trees and grasses, the fertilising pollen is carried by the wind, but this is a somewhat haphazard and wasteful kind of pollination. In a few plants, like the common pea, the pollen of one blossom passes to the stigma of the same blossom, and what is called self-fertilisation occurs. In three or four common flowers, like the dandelion, the egg-cell develops without being fertilised at all—a very rare state of affairs called parthenogenesis. In the great majority of flowering plants the pollination is effected by bees, butterflies, and other insects, which come to the flowers for food (nectar and pollen), and unconsciously carry the pollen from blossom to blossom, thus effecting cross-fertilisation. Thus the question rises: What is the advantage of the insect method of pollination? The answer is that the cross-fertilisation—by pollen from one blossom reaching the stigma of another blossom of the same species—results in more seeds and better seeds than if there were self-fertilisation. Seeds that have two parents are sometimes more vigorous than those that have only one; and seeds that have two parents are more likely to show some new peculiarity or variation, which may sometimes lead to a useful change in the young plant.

If we watch hive-bees at their work, so important for them and so important for the next year's crop of plants, we may be able to verify the fact that on the whole they keep for a while to the same kind of blossom. Aristotle noticed this more than two thousand years ago, and it is very important since it tends to prevent a higgledy-piggledy mixing of the different kinds of pollen-grains that are carried on the hairs of the bee's body and legs.

As we tear ourselves away from watching the summer-bees visiting the flowers in the meadow, two sets of facts rise in our mind. The bees visit the flowers for the sake of the nectar and the pollen, both of which are used as food; yet only a small fraction of the worker-bee's industry is for herself. Workers are females whose development is arrested so that they do not usually lay eggs, and if they lay eggs, these are never fertilised and always develop into drones. Thus we have to face the remarkable fact that this supreme example of summer industry is not entirely for the workers themselves, and that they have not, except in rare cases, any progeny. The marvellous industry is the outcome of an inborn compulsion or instinct to work for the community. Of course the busy bees must have their meals, but the bulk of the food they carry home is for the communal store. Before hive-bees were domesticated the store was used by the bees themselves during the winter months; and so it is in those wild bees that are able to survive the winter as communities. Perhaps the most beautiful form of wealth in the world is the honey in the honeycomb. This self-subordinating industry on the part of the worker-bees wins our admiration, but it has its seamy side. The busy bee works far too hard and rests far too little, and the result is that its brain becomes over-fatigued, as has been microscopically proved, and goes steadily out of gear. With all its getting it gets not wisdom, but premature old age; and an ordinary summer-bee has a short life of six weeks or so, of which about a month may by spent in foraging out of doors.

The other set of facts which the industry of the summer-bees recalls is the attraction that the flowers offer. Many careful experiments have shown that after a bee has discovered nectar or pollen—it usually forages for the one or for the other, not for both at once—it associates its treasure-trove with the particular fragrance of the particular blossom. Hive-bees have an exquisite sense of smell, located in sensitive hairs on various parts of its body, especially on the feelers or antennæ. Having found a blossom that rewards them they try others with the same odour, and when a successful bee returns to the hive and gives over her nectar, she indulges in a brief excited dance on the honeycomb, during which the bystanders rush forward to nose her, thus discovering a clue to the kind of blossom that it would be profitable for them to search for. With this clue they fly off, and it has been shown that in a few minutes they may have discovered the very patch of flowers where the first bee filled her honey-sac. Bees also notice the forms and the brilliance of certain blossoms that are profitable, and they can also distinguish certain colours as colours, though they seem to mix up those that are near one another, such as yellow and orange, or blue and violet. As we are speaking of colour-sense, we may notice that bees seem to be colour-blind to scarlet; but they can see the ultra-violet rays that are invisible to us.

Perhaps we have lingered too long over the bees, but our excuse is that they afford the finest illustrations of summer industry, though not without its tragic note. Thus all the male or drone hive-bees are fatally cold-shouldered at the end of summer, and of the humble-bee community only the young queens survive to see another year. But what are the most important features of the biology of summer? Summer is the season of intensest life, swaying on the pivots of love and hunger. In summer the circulation of matter attains its maximum rate. In spring, both among plants and animals, there is much living on the strength of the past; in summer there is living on the present for the present *but also for the future.*

CHAPTER XIV

THE MARCH OF THE SEASONS: AUTUMN AND WINTER

AUTUMN marks the turn of the tide that began to flow in spring and reached high-water mark in summer. The days become shorter and colder, and there are many storms. Let us repeat: winter is for rest; spring for leafage and new beginnings; summer for flowers and intense industry; autumn for fruits and preparation. We may begin with fruiting and the scattering of seeds, taking them together as characteristic of autumn.

Man does most of his sowing in spring, Nature in autumn; and the chief use of fruits is to protect the seeds—already young lives, as we have noticed—and to secure their scattering. In autumn we gather apples and pears beneath the trees in the orchard—fruits that have *fallen;* as we sit in the sunshine beside a clump of broom-bushes we hear at intervals sounds like little pop-guns, the bursting of the broom-pods, and if we have keen eyes we may see the seeds being catapulted into the air from out of their "legumes"—fruits that *explode.* In autumn we see the dandelion-down and thistle-down with silky hairs, borne like gossamer in the breeze,—fruits that *travel;* we may see the blackbirds feeding on rowan-berries and the like—fruits that are *devoured.* In all these cases the fruit helps in the sowing of the seeds. It is necessary to mention here the rather technical point that when the walls of a dry fruit fit the seed very closely, as in a hazel nut, or in dandelion-down, or in a grain of wheat, fruit and seed are practically the same. Sometimes, moreover, a dry fruit, like that of hemlock, dead-mettle, or mallow, may break into several pieces, each containing a single seed. So that what is sown in these cases is a fraction of the fruit containing a seed; and the same is true when the "stone" of a juicy fruit like a plum is dropped into the ground. We simply mention a point like this because it makes for clear-headedness. By a fruit is meant a ripe seed-box or a number of seed-boxes, sometimes with extra parts added on, as in Cape Gooseberries; by a seed is meant an embryo plant in a resting state; botanically the two are quite distinct, but in some instances, like grains of wheat, they may be practically the same. Dry fruits secure the scattering of the seeds by breakage, sometimes gentle and sometimes violent. In the Sand-box tree, Hura crepitans, the fruit bursts with a noise like a pistol shot, and hurls out the seeds to a distance of many yards. The dry fruits may also float about in the air, like thistle-down; or may adhere to passing animals, like those of "cleavers" or Jack-run-the-hedge in the rabbit's fur. It may be recalled that from one big clod on the foot of a Red-legged partridge, Darwin got eighty seeds to sprout—mostly grasses. Another way in which dry fruits are spread about is by runlets of water which carry the fruits or seeds into streams. It is interesting to notice how cultivated plants sometimes make their way along the banks of a little stream that flows past a cottage garden.

Soft or succulent fruits include (*a*) the stone-fruits, where the innermost envelope of the fruit is very hard and encloses the seed, as in cherries and plums, and (*b*) the berries, where the seeds are directly imbedded in pulp, as in gooseberries and grapes. In both cases the scattering of the seed is secured by the fruit being eaten by animals, especially by birds. This may seem a strange way of sowing, until we notice that only the soft pulp is digested, while the hard seed-containing stone or the hard seeds themselves may be passed out of the food-canal without being in any way harmed. Thus it often pays the plant or the race of plants to have the fruits devoured.

When we look at apples, cherries, oranges, plums, and other delectable fruits, we cannot but be struck by their beauty; and if we are venturesome enough to ask after the *use* of this beauty, the answer must be that the colours, the forms, and the textures of soft fruits serve as advertisements which impress on the bird's mind the fact that these bright objects are very good to eat. Although many birds are partially colour-blind, especially to greens, blues, and violets, there are many that are strongly attracted by reds, which is the commonest fruit colour.

PLATE IV

A COVEY OF PARTRIDGES IN THE HARVEST FIELD

If we raise another question: How fruits come to be so beautiful, we get another kind of answer that lets us into part of the secret of autumn. The beauty of the soft fruit is partly due to its tenseness with syrup, and every green plant is a sugar factory. Moreover, the overflow of sugar from the nectaries of many flowers ceases in autumn, when withering begins, and this closing of the nectaries must leave more sugar available for the swelling fruits. There is also more water available, for the foliage is no longer making such exorbitant demands on the supply. Thus the fruits swell and become beautifully translucent. The question of colour is always difficult, but it should be noticed that one of the common fruit-pigments is anthocyan, which is also common in flowers; and the chemists tell us that anthocyan is a "glucoside," that is to say a compound of sugar and the astringent substance called tannin, which is very common in many parts of plants. Thus it is plain that the raw materials for the coloration are ready at hand. And so, having understood a little of the way in which fruits naturally come to be beautiful, we can with a clearer intellectual conscience return to the previous point, that their beauty has its usefulness in attracting birds.

Before we leave this line of thought we must notice that the exploding of a dry seed-box should be brought into line with the withering, breakage, and fall of the leaf. For a seed-box consists of transformed leaf-structures called carpels, which bear the possible seeds with their hidden egg-cells; and if a dry fruit is composed of leaf-structures, it is natural enough that there should be splitting and bursting in autumn. Thus the exploding broom-pods are to be linked to the falling leaves.

Finally, in connection with fruits, one of the big facts is that if they contain stores, these are mainly carbohydrates (like sugar and starch), whereas seeds are rich in the more complex and, so to speak, more expensive proteins. Part of the meaning of this is plain, namely that what is stowed away in the fruits is lost to the plant race, whereas what is packed into the seeds is a nutritive legacy giving the young plant some capital to start with in the adventure of life. Practically it is a familiar fact that one has to eat a large quantity of fruit to get as much nourishment as there is in a handful of peas, which are seeds.

The drifting fruits of thistles and dandelions and Traveller's Joy or Old Man's Beard recall another of the common autumnal sights—the flight of gossamer spiders, which are scattered by the wind as if they were seeds. When the silken threads, one to four in number for each aeronaut, sink to the ground in thousands we see a " shower of gossamer," often a very beautiful sight. The scattering is of value in enabling the spiders to shift from a crowded to a less crowded haunt, and it is a fine instance of the way in which animals attempt the next to the impossible and achieve it. For who would expect thoroughly terrestrial animals, without wings, to make successful journeys through the air ? Gossamer flights are not restricted to autumn, but they are commonest at this season ; and it must be recognised that there is no hard and fast limitation of particular kinds of activity to particular times of year. Thus summer is the season of greatest animal industry, but earthworms are busiest in autumn.

The aërial journeys of many different kinds of young and small spiders, belonging to species that are fond of the light, are dispersal movements, not true migrations, such as birds so well illustrate. A true migration is a periodic mass-movement from a breeding place to some other haunt which has distinct advantages for the season, and it is followed in most cases by a return journey the following year. It is like a tide between a breeding and nesting place on the one hand, and a feeding and resting place on the other. Now one of the features of late summer and early autumn is the southward migration of birds that have been with us all the summer. The rule is that migrant birds nest in the colder part of their migrational range, and in North Temperate countries it is characteristic of autumn that they make their way, on the whole southwards, to more genial lands, where they recuperate during the winter months. In the summer quarters food is becoming scarcer, for the seeds have been scattered, the fruits have been eaten, and the insects have died or gone into hiding ; the day is much shorter, which makes the quest for food still more difficult ; the cold is setting in and there are threatening storms. Thus it is natural enough that most of the North Temperate birds should shift from their summer quarters and make for the south, often with a swerve to east or west. But it must not be supposed that the external changes do more than pull the trigger of a racial custom that has been in the course of ages somehow engrained in the bird's constitution. Except in the peculiar case of the cuckoo, the young birds leave first, and they have had no experience of any winter. As the poet says, migrant birds "know no winter in their year." The yearlings, who never left their birthplace before, set off on a long journey to an unknown goal; and while there are many tragedies, obedience to the inborn prompting is on the whole rewarded with success.

The word migration should not be blunted by being applied (*a*) to the trekking of lemmings or the swarming of locusts, when the food-supplies are insufficient for the greatly increased population; or (*b*) to the way in which many fishes, like herring and mackerel, follow their food and search for comfortable waters; or (*c*) to a change of habitat with age, as when the delicate larvæ of shore animals swim out into the open sea, and return after they have gained strength. Yet there are not a few examples of true migration in various classes of animals besides birds. Thus there are true migrants among seals and whales, among turtles and toads, in eels (where there is death after reproduction and therefore no return-journey on the part of the adults), in salmon and flounders, in sea-lampreys and in land-crabs, and in a number of other cases. Thus in the Newfoundland caribou, a species of reindeer (Rangifer terranovæ), there is a regular autumnal migration from the storm-vexed mountains of the northern peninsula to the milder regions on the south coast. About the month of October, after the breeding season, when stormy weather sets in and the snows bury the food, the northern reindeer begin to make for the south, travelling mostly by day, usually in small companies and in single file. In the companies in front there is a preponderance of fawns, does, and young deer; in the rearguard, stepping haughtily, come the full-grown stags. If the storms press, the leisurely trek may become a furious stampede. The same route may be followed year after year and become a well-marked road. In spring there is a return-journey, the males leading, the does heavy with young. Sometimes, however, the migration is less well-marked, for according to the weather conditions, many caribou may remain in the north in winter or in the south in summer.

Very characteristic of autumn is the withering and the fall of the leaves. They have worked hard all the summer, and must suffer in some measure from the wear and tear which all activity involves. So they incline to die, except in the hardy evergreens where the leaves last for several seasons, and the fall is often so gradual that it passes unnoticed. In the case of ordinary leaves it is in some ways profitable that they should drop off, for, with their large surface and delicate structure, they would be sources of weakness in winter when soil-water is less available, and when freezing of tissues is apt to occur. Thus it is not wastage that we see when the leaves are shed in autumn, which is often appropriately called "the fall." It is also interesting to find that before the leaves are separated off by a partition, which heals the wound, they surrender to the branch or stem almost all that they have that is worth having. There is a gradual migration of useful substances like sugar. Thus many of the fallen leaves, which the earthworms bury, contain not much more than dead tissue and waste-products.

How transfigured they are in their dying, for some of the woods are so gorgeous in their autumn colouring that we have ventured to speak of the withering leaves as the "flowers of the forest." The yellow, orange, and gold colours are mainly due to the migration or breaking down of the two chlorophyll-greens, so that the two chlorophyll-yellows are left,—for the green of the leaf is a complex of four pigments. In the red, crimson, and purple withered leaves there is often the pigment anthocyan, which we have already noticed in flowers and fruits.

Earthworms are busy all the year round except when the frost grips the earth very hard with its long fingers, but their industry is most marked in autumn. We know this from the large number of "castings" on the meadows, lawns, and putting-greens, and by the twisting trails on the footpath which show how they wandered during the night. What Gilbert White had foreseen in 1777, that earthworms are the most useful animals in the world, was proved convincingly by Darwin in a book "The Formation of Vegetable Mould" which he began as a student of seventeen and did not publish till 1881, the year before he died—a fine instance of his magnificent patience. Darwin showed that there are on an average 53,000 earthworms in an acre of arable ground in England, that they pass ten tons of soil per acre per annum through their food-canals, and that they bury the surface with their castings of finely bruised earth at the rate of three inches in fifteen years. As we see particularly well in the autumn, they drag fallen leaves into their burrows, partly making these more comfortable and partly using the decayed foliage as food. In one earthworm's burrow we found ninety-one leaflets of the rowan-tree; and this burying is going on over vast areas of the world where fertile soil continues to be formed. Ploughers before the plough, the earthworms keep the surface soil circulating, and their burrows allow the air and rain to enter, besides making convenient pathways for the probing tips of rootlets. There is no doubt that earthworms sow many seeds, even of trees.

It seems that earthworms must have originated from a fresh-water stock, for not only do many of their relatives live in ponds and streams, but there are three or four kinds, like Alma and Dero, that have little gills on the sides of their neck. Perhaps the pools became overcrowded, perhaps they dried up too often; in any case the ancestral earthworms began to explore. Thus they discovered the underworld, where they learned to feed on rotting vegetable matter. For a time they must have had a Golden Age, safe from enemies in their subterranean retreats, but by and by their example was followed and the stern struggle for existence began afresh. They were followed underground by poisonous centipedes, who still remain inveterate enemies; by carnivorous beetles; by an occasional flesh-eating slug, like Testacella; and long afterwards by the moles. When they showed themselves above ground, there were birds on the watch, so that earthworms, once so safe, are now among the most persecuted of animals. Thus we can better understand their extreme shyness of the light, their nocturnal prowling, their exquisite sensitiveness to vibrations in the soil, the very rapid reflex action by which they jerk themselves back into their burrow, and even their power of re-growing a new tail, sometimes a new head, if they are cleanly cut by a bird's bill.

We must not linger longer over the Natural History of autumn. It is the season of fruiting and seed-scattering, of withering and retrenchment, of storing and preparation, of showers of gossamer and clouds of migrating birds seeking "warmer lands and coasts that keep the sun."

Winter

It is probable that animals belonging to Temperate, especially North Temperate, regions have been more progressive than those in the Tropics. For there can be no doubt that the punctuation of life by the seasons has been a factor in Organic Evolution. This holds even for winter, though this is very markedly a time of pruning and thinning. Winter often kills, but it also brings the advantage of an enforced rest. It is often a cruel check on life, but it also gives the creature a chance to crouch for another bound—*reculer pour mieux sauter*.

What does winter mean? The days are short, and life usually does best in abundant light; the temperature is low, and only the birds and the mammals have the secret of being able to keep up a constant inside warmth, no matter how cold it may be in the outer world; there are often fierce winds and smothering snow-storms; there is apt to be great scarcity of food. Darkness, frost, storms, and hunger,—these are the sharp pruning-hooks of winter. And just as there were Ice Ages when the greater part of Great Britain, for instance, was covered with glaciers, and most of the animals had to perish or to retreat to the Continent, of which Britain was then an outlying peninsula, so the winters are often hard enough still to prune the tree of life very severely. The Natural History problem of winter is to keep alive, and it is often a pressing problem for mankind as well.

One of the interesting solutions, which only a few mammals have hit upon, is hibernation. This is a better term than "winter-sleep," for the state is very different from sleep, and it may occur in torrid zones at the height of summer, being then called "æstivation." What is this hibernation? It is illustrated by hedgehog, dormouse, marmot, bats, and some other mammals, and when we ask what they have in common, we find that they are imperfectly warm-blooded. That implies a constitutional handicap compared with other mammals; it means that they cannot in the autumn produce enough of heat inside their bodies to balance what they are losing from their skin and in their hot breath. So they sink into a kind of lethargy with feebly-beating heart, breathing movements scarce perceptible, the windows and doors of the senses closely

PLATE V

MOUNTAIN HARE AND PTARMIGAN IN WHITE WINTER DRESS AMID THE SNOW

shut, and the getting rid of poisonous waste-products at a standstill. They sink into a state of suspended animation, somewhat like that of the tortoise, slow-worms, frogs, snails, and humble-bees. Yet it is a peculiar state by itself, and it is bound up with an instinct to creep into some secluded corner or snug retreat, where the temperature rises above that of the surrounding world. If the hibernators "fell asleep" in the open, they would freeze, and a higher animal never recovers if its blood ceases to flow. But inside an old tree, or in a little cave among the rocks, or even in the thickly curtained recesses of the hedgerow, the hibernator may be safe. Their income is nil, and their only expenditure is keeping the fire of life from going out. The state of the blood changes notably and the usual answers-back that the body gives to provocations (reactions to stimuli) are often conspicuous by their absence. Thus a hedgehog may be immersed in a pail of icy water for a quarter of an hour without "wakening" and without being drowned. In most cases, however, the "winter-sleepers" arouse themselves in spring, much the better of their long rest. Out of an undoubted weakness, the hibernators have made a strength!

Along with true hibernation, as we have said, there must be included other forms of suspended animation in lower animals. Of the wasp community, so populous in summer inside their paper-house, only the young queens are left surviving, and they spend the winter, each by herself, in comfortable crannies under bark and thatch and the like. Of the interesting humble-bee community, again, only the young queens are left surviving, and they pass the winter in a sort of sleep, each huddled up by herself, in deep holes in a dry mossy bank. So with the frogs in the mud and the snails in the recesses of the wall, and with scores of other animals. In various states, not very well understood, they lie low and say nothing.

We have spoken of the fall of the leaf, which lessens the exposed and vulnerable surface of the trees, and there are many similar instances of retrenchment as the winter draws near. All over certain grassy hills we see the withering fronds of the bracken, so troublesome in smothering good pasture. One of the reasons why this fern is so difficult to eradicate is just that it retreats in winter into the spreading underground stems or rhizomes which cannot be harmed by the frost. So is it with the numerous plants that have underground corms and bulbs and tubers. Sometimes the whole plant is sacrificed except the root. Thus in the barren grounds or tundra of the Far North, there are vast tracts which are snow-covered and frost-bound for more than half the year ; yet when spring comes the icy desert blossoms like the rose, because there has been life-saving retrenchment beneath the ground.

The same kind of winter retreat is also illustrated among animals. Thus many of the plant-like zoophytes of the shore-pools "die down" in autumn, reminding one of herbaceous plants. The green fresh-water sponge dies away when the cold weather begins to set in, but in the moribund body there are formed numerous, yellowish, pinhead-like clusters of cells called gemmules which eventually float away from the dead skeleton of the parent, and start new sponges in spring. We wish to link this kind of retrenchment to the falling leaves on the one hand, and, on the other, to the mortality in the communities of wasps and humble-bees, where all die except the young queens of the year. For in all these cases we hear the same note of sacrificing parts to save the life. Before the difficulties of winter set in, there is a lessening of what may be a source of weakness. The great army of life becomes entrenched in winter quarters.

Very interesting is the winter whiteness of some mammals, like the stoat and the mountain-hare, the Hudson's Bay lemming and the Arctic fox. It is also well seen in the ptarmigan, the mountain-cousin of the willow-grouse. This bird moults its feathers three times in the year, and the new plumage which it puts on in winter is mostly snow-white. That is to say, the parts of the feathers which contain pigment-granules in the other two plumages are without pigment, but have instead little vacuoles of gas. From the surfaces of these little mirroring bubbles, there is a complete reflection of the light, and that is what we mean by whiteness.

The same is true of the colour-change in mammals, as when the chestnut summer stoat changes into the winter ermine beautifully snow-white, except the black tip of its tail. Sometimes an individual hair changes from brown into white, and intermediate stages can be found ; but usually there is a moulting of the pigmented hairs and a new growth of unpigmented hairs containing numerous minute gas-vacuoles which reflect practically the whole of the light-rays. When an individual hair changes to white, as happens occasionally when a man gets a severe shock, the pigment is partly masked by vacuoles and partly removed by the activity of wandering cells called phagocytes. The Russian biologist Metchnikoff studied the process of rapid blanching, and found that the mobile devouring-cells or phagocytes (occurring in most animals) pass up into the hair or feather and return into the skin with a microscopically minute burden of pigment. He was able to see different stages of the process in different hairs. When a man's hair grows grey slowly, and when there is hair-cutting, the process is usually that the ends are cut off, while the new growth at the base has less and less pigment as the years go by.

But let us return to the seasonal colour-change in such mammals as the stoat and the mountain-hare, where there is a moult and a new coat ; what is the advantage of this ? It is often said that the white mountain-hare is hidden against a background of snow from the hungry eyes of the fox and the golden eagle ; and there may be something in this interpretation of the whiteness as a cloak of invisibility. Our suspicions are aroused, however, by the fact that when there is most snow on the hills, the hares come off them in large numbers in search of food at lower altitudes. But against the bare patches they are exceedingly conspicuous, challenging attention. Similarly in regard to the ermine, it has few enemies that it need hide from, and if we say that its whiteness enables it to steal upon the unsuspecting ptarmigan amidst the snow, we must in fairness remember that the ptarmigan has also turned white. We do not wish to deny the possibility of there being an advantage in the cloak of invisibility, but it looks as if this was not the whole story. What other advantage can there be in a suit of white fur or white feathers ?

The answer is not far to seek and it applies also to those mammals like the polar bear, and to those birds like the snowy owl, which are permanently white and live in the Far North. The answer is that for a warm-blooded animal in very cold surroundings the most profitable colour is white. Less heat is lost from white fur or white feathers than from any other colour, and the advantage of conserving the precious animal heat is great. It economises the energy of the body not to produce more heat than is necessary, and the animal heat makes it possible for the chemical routine of the body to proceed more rapidly and smoothly.

FIG. 14.—SNOWY OWL, *NYCTEA NYCTEA*.

A circumpolar Arctic bird, adapted to snow-covered surroundings, an occasional winter visitor to the North of Scotland.

How do animals meet the winter? That was our question. We have noticed some of the ways—by migrating, by hibernating, by retrenchment, by storing, by blanching, and so forth. Yet after all we must not try to hide from ourselves the fact that winter is also the season of sifting. Many migrants never return, many winter-sleepers never re-awaken, many retrenchments are all in vain, and many stores are stolen by enemies who break through. Winter is a time of pruning and sifting, and can we say anything better than what Goethe said, that "death is Nature's device for securing abundance of life." Stern winters mean finer springs.

CHAPTER XV

THE EVERYDAY LIFE OF THE BODY

WE have pictured the peopling of earth and sea and air and also the changes in the aspects of life from season to season; it is now time to turn to the living creatures themselves and to ask, first of all, what goes on in the animal body. This is PHYSIOLOGY.

The business of living creatures has mainly to do with caring for self and caring for others. Caring for self includes first of all the quest for food and other necessaries like fresh air and water; it also includes all forms of self-preservation, such as avoiding enemies and keeping a firm foothold amid changeful surroundings. Caring for others

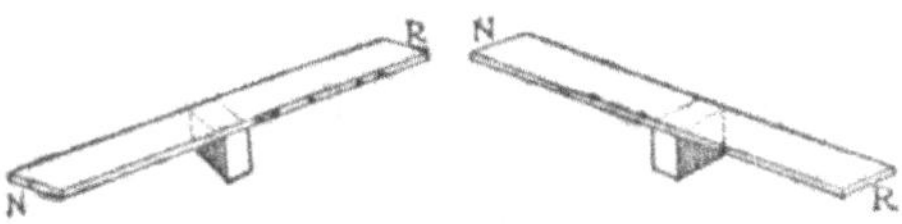

FIG. 15.

The frequent alternation and antithesis between nutritive processes (N) and reproductive processes (R).

includes the love of mates and the upbringing of offspring, and in some cases, such as ants and bees, the service of the community. The poet asked: "Why do the people so strive and cry?" And he gave the answer, true for animals as well as for men: "They will have food and they will have children, and they will bring them up as well as they can." As another wise saying puts it: "While philosophers are disputing, hunger and love solve the world's problems." For in "hunger" and "love," looked at broadly, we see the twofold business of human life, and of animal life as well.

But our question is not concerned with the deep impulses of life; it has to do with the everyday bodily activities to which the impulses lead. Hunger is an appetite, and food is an end in view, but what are the means? Love is the deepest of feelings, and it finds what it seeks in mates and offspring and kindred, but how does the creature keep agoing from day to day? What is always going on in the living body, and how does it keep agoing? What corresponds to the burning of fuel and the whirring of wheels in the engines that work for man? Let us think of what have been called *the engines of the living body*, though more than engines they certainly are.

(1) The first place must be given to the locomotor engines of the body—the muscles, which usually work by pulling one piece of skeleton nearer another, thus levering the animal along. They are "pull-engines," whereas a motor bicycle or any similar contrivance is a "push engine." A piece of muscle or flesh is made up of numerous parallel threads or fibres, really transparent when they are living, but often appearing reddish because of numerous minute blood-vessels. Their transparency is clearer in the flesh of lobsters, where the blood is not red, or in the flesh of fishes, where there is less blood than in mammals. Each muscle-fibre consists of still finer invisible filaments—the fibrils, and hundreds of fibres are wrapped up together in a sheath of shining connective tissue to form a muscle, which usually runs from one piece of skeleton to another. The sheath of a muscle is continued in backboned animals to form a strong tendon or sinew, and this is attached to a bone. Of course there are some muscles that are not connected with the skeleton, such as those that make the heart beat and those that force the food down the food-canal. Such muscles by their contraction lessen the size of the cavity which they enclose, thus driving on the contents, quickly in the case of the blood from the heart, slowly in the case of the food that is being digested in the food-canal.

When we draw our lower arm up towards our shoulder, we are using a muscle called the biceps, which has about half a million fibres. If we put the fingers of our right hand on the upper left arm when we are contracting our biceps, we feel the flesh rising. The muscle is becoming shorter and broader, as a whole and in all its fibres; and it does so in obedience to a command that comes by a nerve to the muscle, by a nerve-fibre to each muscle-fibre.

What happens when a muscle-fibre receives its orders from a nerve-fibre? This is a question very difficult to answer, but we may say that there is, to begin with, an instantaneous production of lactic acid which bathes each fibre. Whereupon, as if a spring were let loose, the fibre contracts. The next chapter is a rapid combustion of part of the lactic acid, and the energy thus produced is utilised to reinstate in the fibre what is left of the lactic acid, so that the muscle may be ready to contract again. In the first chapter, which is physical, there is an electrical discharge; in the second chapter, which is chemical, there is a production of heat and carbonic acid gas.

(2) If the animal is to move about in an effective way, it must keep in touch with the surroundings and it must control its muscular contractions. This is the twofold task of the nervous system, and long before the brain of animals was used for thinking, it was the governor of movement. Similarly it must be said that motion is much older than emotion.

In all animals from earthworms to man the nervous system includes three different kinds of nerve-cells, connected to one another by fibres. First there are the sensory nerve-cells which receive tidings or stimuli from the outer world, and are often compacted together to form sense-organs, such as eyes. On certain patches inside our nostrils there are groups of sensory nerve-cells which are affected by odours, and send their news by the olfactory nerve to the front of the brain. Similarly, in backboneless animals the sensory nerve-cells are usually on the surface of the body. In backboned animals, however, except in the case of the sense-organs of the head, the sensory nerve-cells are very deeply situated in the spinal ganglia, which lie close beside the nerve cord or spinal cord. They form the spinal ganglia of the dorsal roots of the spinal nerves that issue from the spinal cord. (See Fig. 16.) These deeply situated sensory nerve-cells give off sensory nerve-fibres which extend to sensitive nerve-endings in the skin. They may end freely or they may be surrounded by special cells forming touch-spots and other sensitive spots. From the sensory nerve-cells in the spinal ganglia other fibres pass into the spinal cord, where each divides up.

If the nerve-fibre that ends on a muscle-fibre be traced backwards, it joins with other motor nerve-fibres to form a motor nerve; and when that is followed it passes into a spinal nerve, and then by the ventral root into the spinal cord. There it breaks up again into nerve-fibres, each one of which leads to its origin—a motor nerve-cell in the spinal cord. It is from these motor nerve-cells that the commands come that prompt a muscle to contract.

There is a third kind of nerve-cell, however, which is called associative or connecting, and serves as a middle-man in the spinal cord between a sensory nerve-fibre from a sensory nerve-cell and a short branch of a motor nerve-

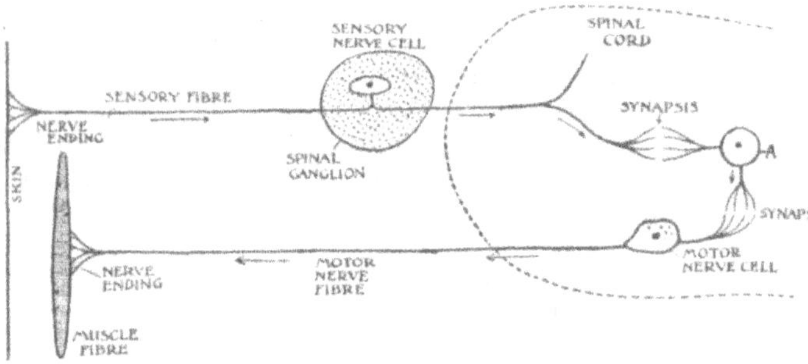

FIG. 16.—TYPICAL REFLEX ARC IN A VERTEBRATE.

The sensory neurons lie in the spinal ganglia just outside the spinal cord. A spinal nerve is made up of sensory nerve-fibres on their way into the cord and of motor nerve-fibres on their way out. A, an associative neuron.

cell. It is not easy to follow this linkage the first time, but with the help of a diagram it soon becomes quite clear as long as we keep to the elementary facts. Let us think over these elementary facts, which are so important that we have emphasised them again in speaking of mind.

The *sensory* nerve-cells are like scouts, but they collect their information not only from the outer world, but from the body itself,—not only from the surrounding country, so to speak, but from the camp. We shall call them S. The associative nerve-cells correspond to General Head-quarters in an army; they receive the tidings which the scouts send in. We shall call them A. The motor nerve-cells correspond to Executive Officers, who send out orders, saying to one "Go," and he goeth; and to another "Come," and he cometh. They are of different ranks, like Majors, Captains, Sergeants, and so forth. We shall call them M. Fourthly there are the muscle-cells or muscle-fibres that do the work, and might be compared to the soldiers who do what has to be done. They are often called "effectors," and we shall use the contraction "E." Thus taking the first letters of the names of the four kinds of cells—Sensory, Associative, Motor, Effector—we have S → A → M → E; and this sequence is what is meant by a "reflex action."

Picture an earthworm half-way out of its burrow; it feels the tremor in the earth caused by the light tread of a blackbird's foot; it jerks itself back into its hole and is

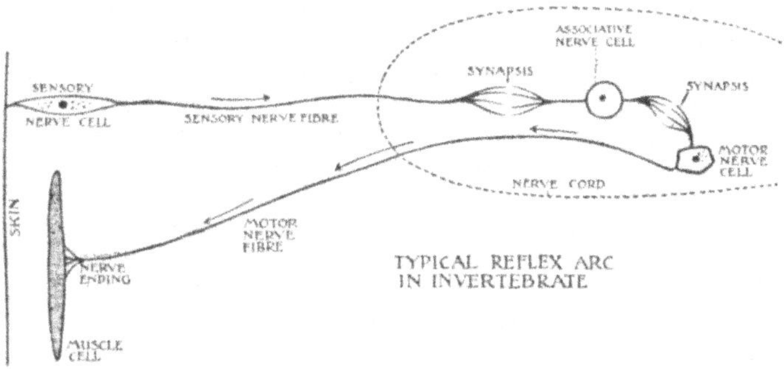

FIG. 17.
As in Fig. 16 the arrows indicate the direction of the nerve impulses.

safe for the time being. It does not think about it, neither does it will to retreat; what happens is a reflex action. Sensory-cells on the skin are stimulated by the slight vibration; a message travels from the fibre of each of them into the nerve cord which runs along the ventral middle line of the earthworm; the thrill passes into associative nerve-cells and thence to motor nerve-cells; from the latter a command is sent to the muscles of the body-wall, which contract,—and the animal is back in its hole. The same kind of reflex action, but differing in details, is illustrated when we draw away our finger from a hot cinder, or close our eye on the approach of a missile, or cough away a crumb from the opening of our windpipe. We must return to this when we come to discuss "mind."

Not very much is yet known as to the nature of the message that travels along a nerve-fibre, from the sense-organs to the central nervous system, or from the centre to the muscles and glands. A common rate in man is about 400 feet in a second; in the frog about a quarter of that; in some backboneless animals much less. There does not seem to be much using up of material when the thrill passes, but some recent investigations have shown that there is a slight production of heat and of carbon dioxide, which points to the occurrence of some chemical reaction. As with most vital phenomena, the excitement of the nerve is associated with an electrical change.

(3) The activities of the animal body, both external and internal, involve expenditure of energy; and the source of that energy is to be found in the burning away of carbon-compounds, just as in a steam-engine or a motor-car. The carbon-compounds arise from the food, which is transformed by digestion; and food is also required to make good the wear and tear, and to provide fresh building material as long as the animal is growing. In an ordinary animal the food passes in solid form into the food-canal and is there digested. This digestion means in part that the solid material becomes liquid, thus the solid starch grains of the potato become liquid sugar; the fatty food is changed into fatty acids and glycerine; the protein, such as the vitellin of yolk of egg, the casein of cheese, or the gluten of wheat, is dissolved into peptones and then broken down into what are called amino-acids. This liquefying allows the dissolved sugars and amino-acids to diffuse through the walls of the food-canal into the blood; and the dissolved fatty acids similarly pass into a system of lymph vessels, which communicate in various ways with the veins, and thus eventually with the heart. Evidently one of the great uses of the blood-stream is to transport the digested food throughout the body, to places where it is required. But there is another and more difficult point. As living matter always consists in great part of proteins in a watery medium, it is not surprising that proteins should form a large part of ordinary food. But proteins are not good mixers, and the introduction of a strange protein into the blood of an animal is apt to be fatal. This is familiarly illustrated by the consequences of snake bite, for the poison of snakes is a protein, which cannot mingle harmoniously with the

proteins of its victim's blood. Here then is another use of digestion, for the proteins of the food are split up into the simpler, harmless, but very useful, amino-acids.

(4) The source of energy in animals is the combustion of carbon-compounds derived from the food. It is therefore necessary that there should be a continual supply of oxygen to keep the fire of life burning. It is also necessary that the waste gas formed in the combustion or oxidation of carbon-compounds should be got rid of, for it is a poison. This then is the great function of respiration,—the supply of oxygen and the removal of carbon dioxide. The place where oxygen is captured in higher animals is the internal surface of the air-filled lungs; and the place where it is used up is in the tissues, like muscles and glands, and including the blood itself. So the tissues produce carbon dioxide and it is got rid of on the internal surface of the lungs. We see, then, that a second great use of the blood is to carry oxygen from the place of capture to the place of usage, and to carry carbon dioxide from the place where it is produced to the place where it is got rid of. The oxygen is captured because the red pigment (hæmoglobin) of the blood has a strong chemical affinity for it, though readily yielding it again to the still stronger chemical claims of the living matter.

(5) If living matter consists in great part of proteins, which are *nitrogenous* carbon-compounds, and if living involves the breaking down of the complex into the simple, then there must be a formation of nitrogenous waste-products. They are also formed from the unused residue of digested nitrogenous food. But nitrogenous waste products are apt to be poisonous, in the form of ammonia for instance, so they must be got rid of, or excreted, as quickly as possible. In higher animals this excretion is begun in the liver and finished in the kidneys; and this is a third great use of the blood, to collect the fluid nitrogenous waste and to transport it to the kidneys where it is filtered out. So we may sum up by saying that the everyday functions of the animal body are—contractility, irritability, digestion, respiration, and excretion.

(6) Besides these there is the regulating of the complicated routine so that it works smoothly. This is the work of the chemical messengers or hormones which are produced by the ductless glands, such as the thyroids which lie on

each side of our larynx or Adam's Apple. The invisible hormones pass into the blood and are distributed throughout the body, where they serve as accelerators and as brakes. Our health of body and mind depends on these regulators which help the different organs of the body to work together harmoniously. This twentieth-century discovery has changed the face of physiology; it has made both the everyday and the extraday life of the body much more intelligible.

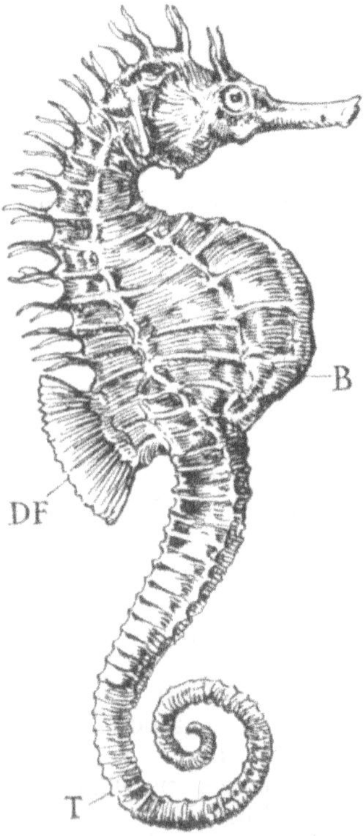

FIG. 18.—MALE SEA-HORSE, *HIPPOCAMPUS*.

DF, rapidly vibrating dorsal fin; B, ventral pouch in which the male fish carries the developing eggs till they hatch; T, prehensile tail, movable dorso-ventrally, often twisted round seaweed.

A living body is in some ways like an engine, for it does work, and the food corresponds to the fuel. But the living creature stokes itself, repairs itself, regulates itself, and sets itself tasks. In many cases among higher animals it is quite clear that the creature has a purpose in its head, and that it can change its mind. Thus the animal is ever so much more than an engine; it is like a combination of an engine and an engineer.

Another outstanding difference is that living creatures give rise to other living creatures like themselves, and no engine ever does anything like that. We mention this because if we inquire into birds or bees, dogs or dragon-flies, fishes or frogs, sea-urchins or sea-anemones, we find that part of the life of the body has to do with producing offspring. In many cases a great part of the animal's energy is devoted to securing the welfare of its family. There is caring for others as well as caring for self. We shall come back to this when we study the life-creature's life-curve or trajectory; but we must first make a comparison between plants and animals.

CHAPTER XVI

THE BEHAVIOUR OF PLANTS

OF recent years the outlook of Science on the world of plants has become much more generous than it used to be. It has been proved that the average plant is nearer the animal, in a sense, than Linnæus thought when he said: "Stones grow; plants grow and live; animals grow and live and feel." If "feel" means answering-back to outside influences, there is no doubt that plants feel. They have a life of far greater intensity than was previously suspected.

Many animals like zoophytes and corals are fixed, and form colonies by budding. A Black Coral is often like a thickly growing bush with delicate twigs, and many of the sedentary animals become like little trees. Hence names like "Sea-Fir." It is interesting to find that the fixed sea-squirts or Ascidians, which are half-way up the scale of animal life, are clothed in tunics of cellulose, which is the characteristic vegetable substance that forms the cell-walls of plants and changes into wood in trees. Thinking of such cases, we may say that there is a good deal of the plant in many an animal. It is equally true that there is a good deal of the animal in many a plant. Think of the sundew catching flies on the moor, the exquisitely "touchy" Sensitive Plant, and those climbers that have mobile tendrils like those of the Bryony and the Vine. Some of the simplest green plants swim about actively, and there are others called Oscillatorias that are always waving to and fro in a somewhat puzzling fashion. There is much mobility and much sensitiveness in the world of plants.

There is no doubt, of course, that plants and animals are on entirely different lines of life. No greater parting-of-the-ways has ever occurred than that which made the genealogical tree of living creatures somewhat V-like, with plants on the one fork and animals on the other. At the short common base of the V there are very simple living creatures, sometimes called Protists, which have not taken a decisive step towards plants or towards animals; and, as we noticed in one of our early studies, it is probable that the first living creatures (or organisms) were not very unlike the simplest of the present-day Protists. But, while they may have had a common origin, plants and animals have diverged in very different directions, so that when we pass to the top-ends of the fork, we are apt to see far more differences than resemblances, say between birds and the trees on which they perch.

The fundamental difference is doubtless that all green plants are able to feed at a low chemical level, on air and on soil-water with its dissolved salts. They are able to do this because their greenish pigments helps them to utilise the energy of some of the red-orange-yellow rays of the sunlight to build up complex carbon-compounds like sugars and proteins. This up-building, to which we have already referred, is the most important process in the world; and it occurs in plants much in excess of their immediate needs. The up-building is far greater than the down-breaking, and thus there is abundant reserve material for storage, on which almost all forms of animal life are sooner or later dependent. The ordinary animal feeds at a high chemical level, for it cannot obtain the requisite carbon from anything simpler than carbohydrates and fats, and it cannot obtain the requisite nitrogen from anything simpler than proteins. It gets these carbon-compounds, directly or indirectly, from plants. With this complex food, rich in potential energy, the animal is able to live a life of great activity; its expenditure is often close up to its income, a striking contrast to plants. One might say that plants manufacture munitions somewhat slowly and carefully, whereas animals take these and explode them, sometimes very lavishly.

But in spite of this deep-down parting-of-the-ways, it has been clearly recognised since the days of the great French physiologist, Claude Bernard, that there are many samenesses in the life of plants and animals. If we study a green plant at night it is not difficult to prove that it is taking in oxygen and liberating carbon dioxide, just as we do constantly in our breathing. If we have in a room at night a burning candle, a green plant, and a purring cat, all three are using up oxygen and returning carbonic acid gas to the air of the room. In other words, the plant shows the function of respiration, but this is masked during sunlight hours by the converse nutritive process, peculiar to green plants, of splitting up the carbonic acid gas and returning the oxygen,—a process to which the earth owes its breathable air.

Similarly, the plant has its digestive ferments, not merely in the exceptional insect-eaters like sundew and Venus's fly-trap, but in all ordinary plants. In every green leaf there is a (diastatic) ferment which changes starch into sugar ; in every sprouting seed there is a (peptic) ferment which changes condensed protein into more available form —mobilising it as one might say.

To an extent that was unexpected at the beginning of this century, plants form nitrogenous waste-products, sometimes even the urea that is so characteristic of mammals. These nitrogenous waste-products are the ashes of the living fires, and they would be as noticeable in plants as in animals, were it not that plants change the poisonous ammoniacal waste into a harmless form, and then, when necessary, use this over again as part of the nitrogenous food. This works very well, for plants are apt to suffer from nitrogen-deficiency when the soil is poor, whereas animals are more apt to suffer from nitrogen-excess. We must not follow this further, for the present point is simply that plants have a function of excretion just as animals have, though it is somewhat disguised, and the same may be said in regard to respiration and digestion. It may also be noted that some plants, e.g. the Sensitive Plant, have hormones, those mysterious self-made medicines which are so indispensable in the internal economy of the higher animals.

The beech-tree feeds and grows, digests and breathes as really as does the squirrel on its branches. Moreover, both start in their individual life from a very minute fertilised egg-cell. Both develop by the oft-repeated dividing and redividing of their cells, and by a puzzling process called division of labour, so that the final result is a vast community, often of many millions, of cells, differing from one another in structure and in function. It is clear, then, that in regard to many fundamental features the plant and the animal are nearer one another than was formerly supposed. But how do plants stand as regards the two *master-activities* of the animal, namely moving and feeling ?

As regards *moving*, the growing stems of plants show a gentle swaying, bending and bowing—" nutating " as it is called—to the different points of the compass. The tip of a root in an earthworm's burrow moves tentatively; leaves rise and fall, flowers open and close, with the waxing and waning light of day, and with the associated changes of temperature. Moving and feeling, we see them both in twiners and tendril bearers, in the leaves and leaflets of the Sensitive Plant, in the tentacles of the sundew and the neatly-working Venus's fly-trap, in the stamens of the barberry and the rock-rose, in the stigma of the musk, and in many other plants. Sir Jagadis Chunder Bose has given his strenuous life to answering such questions as these: Are all plants sensitive to stimulus ? Is there, as in animals, a pause or latent period between receiving the stimulus and giving an answer ? The answer to both these questions is an emphatic " Yes." He has gone on to ask whether an impulse can spread through a plant like the nerve-impulse in an animal; and again the affirmative answer is emphatic, that whatever the interpretation may be, there is no doubt that a message can travel from one part of a plant to another, and that plants may be chloroformed, fatigued, and poisoned much in the same way as animals. By attaching delicate and ingenious recording instruments to the plants experimented with, Bose has been able to show how their growth waxes and wanes with the weather, and how their answers-back vary in promptness with the external conditions. He has got the plants to write what he calls " autographs," which give some indication of their changing states. He has made " the dumb plant the most eloquent chronicler of its inner life and experience." After many years of patient and clever experimenting, he has been led to the daring conclusion that " there is no life-reaction in even the highest animal which has not been adumbrated in the plant."

The distinguished Indian physiologist is only one of many investigators who have revealed the sensitiveness and mobility of plants, and while it is not necessary to accept *all* his conclusions, such as that there are lines of cells in the Sensitive Plant that correspond to nerves in animals, we take his work as a fine instance of the new outlook and of extraordinarily ingenious experimenting. In any case, Bose has proved that every plant is a "Sensitive Plant." Personally we must confess that we do not think that he has proved that any plants have nerves or a heart.

By attaching a delicate recorder to the Sensitive Plant proper, Mimosa pudica, which is called in Bengal "the coy maiden," it is possible to study its responses with great precision. In answer to a sharp blow or to the touch of a glowing incense stick, the numerous leaflets of the beautiful compound pinnate leaf close together from below upwards; then the four ribs draw together sideways as in a folding fan; then the leaf as a whole sinks down against the branch, moving on the base of the leaf-stalk. But more than that, the impulse passes into the stem and other leaves collapse. If the provocation or stimulus be strong enough, the impulse, whatever it may be, ascends to the apex of the stem and travels down the other side—a striking observation which seems to have been missed by previous investigators. It seems to Bose that a nervous thrill or impulse passes along certain lines of cells from leaflet to rib, and from rib to leaf-stalk, and onwards; but according to the researches of Ricca and of Snow what happens is a rapid diffusion of a "chemical messenger" or hormone. This shows how equally clever investigators may come *for a time* to different conclusions. But in any case Bose has improved on the distinction which Linnæus drew between plants and animals. We may now say: Vegetabilia crescunt et vivunt et sentiunt: Plants grow and live and feel.

The movements of the Sensitive Plant are due to changes in the internal pressure (or turgidity) of certain cells in sensitive cushions which occur where the leaflets join the ribs, and where the four ribs join the main leaf-stalk, and where the leaf-stalk joins the branch. The upper surface of each cushion is less active, less sensitive, less contractile, than the lower surface. Therefore when the stimulus comes, there is inequality of response on the two surfaces, the result being a definite movement. The same is true in regard to the two surfaces of tendrils, which are so sensitive that they answer back to the touch of a stretched thread of silk.

In an animal an appreciable time elapses between the

stimulus and the responsive movement. It is about a hundredth of a second for a frog, but much longer for the sluggish tortoise. This "latent period" is also observed in plants, but it may be only 0·075 of a second. It is lengthened out in an interesting way by fatigue, or by a chill, or by a darkening of the sky. Since the life of plants is so closely bound up with light, it is not surprising that they should be many times more sensitive than man to changes in illumination. On one occasion Sir Jagadis noticed a

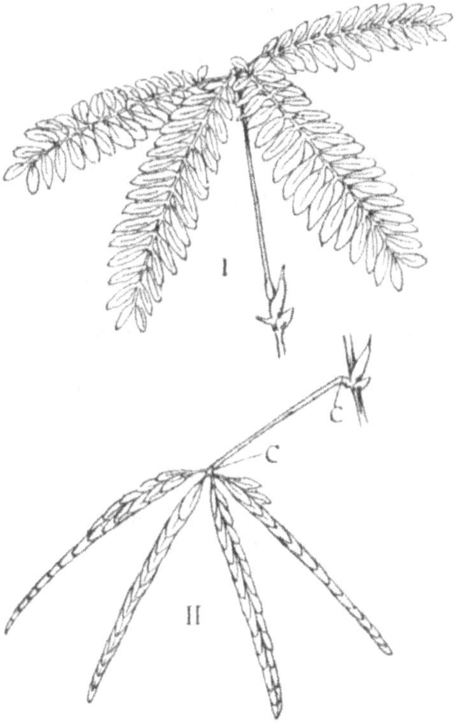

FIG. 19.—TWO LEAVES OF THE SENSITIVE PLANT, *MIMOSA PUDICA*.

I expanded, II folded together and sunk downwards. The positions of the two largest cushions (C) are shown at the base of the main leaf-stalk, and at the bases of the midribs of the four pinnæ which bear the leaflets.

sudden sluggishness in the response of the plant that he was working with ; it puzzled him, for he could detect no change in the conditions of the laboratory. But on looking through the window he noticed that a wisp of cloud was passing across the sun !

One of the most sensitive parts of our body is the tip of the tongue. If a silver and a copper coin be held in the mouth in contact with the upper and under surface of the tongue, there is a slight electric discharge whenever the two coins touch. We notice a slightly acid taste. To a weak discharge of this kind the Hindu student is twice as sensitive as a European, but the common weed called Biophytum is four times more sensitive than the Hindu !

If a Venus's fly-trap (Dionæa) is cheated two or three times with faked " fly " instead of getting real fly, the trap ceases to work. Yet in a short time it has " forgotten " and may be cheated once more. Its " memory," if one dare use the word, is short, yet there is evidently some enregistering of experience so that subsequent behaviour is influenced ; and there are other cases of the same sort. This is probably as near behaviour as a plant has got ; and it is just possible that there is here a first awakening of the vegetable " mind." But we should not pay any serious attention to people who tell us that a cabbage suffers when it is pulled out of the ground. They say that they have heard it squeal a little when it is roughly handled.

CHAPTER XVII

THE BEHAVIOUR OF ANIMALS

IN our fifteenth study we asked how the living body works, and how it keeps agoing from day to day. The answer to that question is ordinary PHYSIOLOGY. Now let us inquire into the doings and ongoings of animals, how they direct their activities towards particular ends. Let us think of them as creatures that do things; let us ask whether they have feelings, thoughts, and wishes as we have. This is the study of Behaviour or PSYCHOLOGY.

Recent observations on the higher apes show that they have a high degree of intelligence. We mean by intelligence some understanding of the relations of things, some spice of judgment, some power of saying "If this, then that!" This is illustrated when chimpanzees pile up four boxes, one on the top of another, to reach a banana hanging from the roof; or when they piece together two lengths of bamboo rod so as to make one rod long enough to reach some fruit outside the cage; or when having discovered the use of a lever or a hand-mirror, they proceed to find other levers and other mirrors—including in the last case a pool of rain-water. It does not seem possible to describe behaviour like this without supposing a mind at work. There is some hint of *thinking*, even if it be no more than a kind of picture logic. There is, within limits, an intelligent grasp of the situation.

This kind of behaviour, which implies an inner life of ideas and purposes, and a making of experiments inside the animal's head first of all, is also clearly seen in some of the other mammals, such as elephants, horses, and dogs, and in some of the birds, such as rooks, cranes, and parrots. It is to be distinguished from the outcome of training where individual habits are sometimes established without the animal being intelligently aware of what it is doing.

Thus the elephant that accepts pennies and puts them into the slot of the biscuit-delivering machine, but rejects halfpennies with impatience, was able to do so as the result of very laborious training. Similarly with most of the feats of the Indian weaver-bird in picking out particular cards or coins.

When a chimpanzee whittles with its teeth at the end of a bamboo rod so as to get it to fit into the end of another piece, thereby making a rod long enough to retrieve a fruit that is out of reach, this is on a much higher level than random "trial and error" experimenting, which is very common among animals. In this simple experimental behaviour many movements are tried one after the other, with the hope no doubt that one may prove effective. Thus one of the broad-nosed monkeys was seen to spend a long time in shaking a narrow-necked bottle in the hope of getting at the pea-nut inside. Eventually it got what it wanted when it happened to hold the bottle vertically inverted. But this did not in itself prove much cleverness on the monkey's part. If it should afterwards skip all the useless movements, having perceived why one only is successful, our estimate of its intelligence would rise, and in proportion to the shortness of the time required in the cutting out of what is useless. But here again we must not be hastily generous, for many an animal has the power of somehow registering *in its body* the movements that proved effective in a trial and error experiment. Even without its head an earthworm in a T-shaped tube will learn almost perfectly to avoid entering that branch of the T that has a mild electric shock at its end. This is effective profiting by experience, yet it is not intelligent.

But we must also try to distinguish intelligent behaviour, sometimes exhibited by the higher mammals and birds, and almost always by ourselves, from *rational* conduct which is only occasionally illustrated in what we do. Animals sometimes reason at the level of picture logic (perceptual inference), as man habitually does, yet they have no Reason, in the strict meaning of that great word. For it is generally agreed that the word Reason should be kept for working with general ideas, which the psychologists call concepts. Our mental picture of a particular horse is a "percept"; our idea of horses in general is a "concept." Now no instance of animal behaviour has been recorded which demands for its satisfactory description that we should credit the creatures with the power of working with general ideas. Even very intelligent animals like chimpanzees soon reach their limit, and are puzzled where a child of the same age would see the way out at once. There seem to be two chief handicaps. First, the chimpanzees have a very limited power of forming and storing mental images, for the ingenious solutions they sometimes discover are almost always, if not always, by means of things that they can see at the time. Second, they have no more than the beginnings of language. For while chimpanzees, to keep to them, have many *words* which are signs for particular things or particular feelings, they are not known to make even the simplest sentences, putting words together so as to express a judgment, such as "Banana good" or "Mamma back soon"—as a child three years of age or less, might say. No doubt they chatter a good deal among themselves, but their chatter is not even gossip; it is little more than a succession of grunts and grumbles. A two-year-old child gets much further.

On a different line from intelligent behaviour is instinctive behaviour, so well illustrated by ants, bees, and wasps. It does not require to be learned, the power is inborn. If a child was able to play the piano the first time it tried, that would be instinct; and that is just the kind of thing a young spider does when it makes its web true to pattern at its first attempt. An instinct is an inborn power of doing apparently clever things, yet this does not mean that it may not be sometimes improved by practice. It may also be helped out by intelligence, as is often illustrated by the ways of birds in nest-building and the like; but there is no reason to believe that in its ordinary routine a bee understands what it is doing. It obeys its engrained impulses, and it is often nonplussed by a very slight departure from normal circumstances. No doubt it is useful that procession caterpillars should persist in going on and on in an Indian file, the head of one touching the tail of the next in front, for if they go on and on they are likely to find a suitable patch of soft soil in which they may bury themselves and undergo the great transformation into winged insects. But when the Italian boy twists the Indian file carefully round till the head of A touches the tail of Z, the caterpillars continue for hours going round and round in a useless circle. Sooner or later a stumble breaks the circuit and breaks the spell; but it is not intelligence that saves the situation. Instinctive behaviour is always related to particular circumstances which pull the trigger of activity, and a slight change often seems to puzzle the animal hopelessly. It is quite different with the intelligent animal, which has some understanding of the

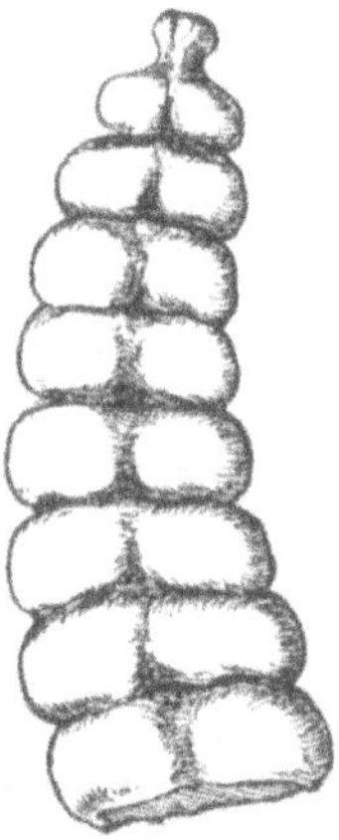

FIG. 20.—RATTLE OF RATTLE-SNAKE, *CROTALUS HORRIDUS.*

When the snake casts its slough, i.e. the outermost layer of the epidermis, extending over the scales, it turns it inside out from the head tailwards. In the Rattle-snake a terminal portion of this catches at the end of the tail, and a loosely jointed horny rattle is built up. The number of rings or bells does not usually exceed twelve, for the older ones break off from time to time. When the snake is excited it rattles its rattle, producing a shrill sound; this serves to warn off large animals on which it would simply waste its poison, and perhaps break its teeth.

situation and can alter its behaviour to suit. Another contrast is that intelligence often differs markedly from individual to individual, whereas instinct is much the same among the members of the same species, or at any rate among those of the same sex. All the female garden spiders make equally perfect webs. It is plain, then, that instinctive behaviour is not a low form of intelligent behaviour, but is on a quite different line.

Instinctive behaviour, looked at from the bodily or physiological side, depends on inborn linkages between certain nerve-cells and certain muscle-cells. Just as we do not *learn* to sneeze or cough, but do so because of certain (reflex) linkages established in our body before birth, so, but in a much more complex way, the bee does not require to learn how to find and how to enter flowers when it takes its first flight in the open air after its apprenticeship to quite different tasks in the semi-darkness of the hive. But in addition to these inborn linkages which lead the creature to follow a certain line of behaviour, it seems necessary to credit the instinctive animal, in some cases at least, with something in the way of impulses and consciousness. In many cases instinctive behaviour seems to be suffused with awareness and backed by endeavour. Its great advantage over intelligent behaviour is that it is, so to speak, ready-made, not requiring any apprenticeship. Its great disadvantage is that it is not plastic; it is not readily adjusted to meet changes in circumstances. At a critical point, however, intelligence sometimes comes to the rescue.

An interesting kind of behaviour is well illustrated by the young European eels or elvers which swim up the rivers in spring. They are already over two and a half years old, and they have had a long sea-journey before they reach the mouths of the rivers. Indeed both the European and the North American eels begin their life in a great stretch of sea not very far from the Bermudas. When they reach the quick-flowing river they usually hug the banks, avoiding the main current; but their inborn impulse is to go straight on as long as the daylight lasts. When the sun sets they snuggle under stones and under the bank and wait till next morning. These elvers, about the length of one's first finger and the thickness of a bone knitting needle, are so constituted that they must keep the two sides of their body equally pressed upon by the water. If they are driven off the straight path by the entrance of a tributary, they automatically right themselves, just as a spinning top or a gyroscope does, only more subtly. It is not that they *try* to get straight again; it is rather that they do this automatically. Their muscles strike the water more forcibly on one side than on the other, until the two sides are equally pressed upon, or, perhaps one should say, not pressed upon, by the current. This kind of movement is called a forced movement, or *tropism*, and it is illustrated by many animals that move towards or away from light, or moisture, or gravity, or certain chemical peculiarities, and so on. It is always towards or away from the *direction* of the stimulus. It often saves the animal much time and effort, and in most cases it works well. But if there is something very unusual in the surroundings, the " forced movement " may work badly. Thus if a moth comes flying past a candle, one eye is bound to be more illumined than the other; the insect's constitution compels it to adjust its wing-strokes so that both eyes are equally illumined. But if it goes on flying it is in this way almost compelled to fly into the flame. Unless it should turn outwards so that both eyes are equally unillumined it can hardly help itself. The candle is not part of a moth's natural environment.

Somewhat in the same line is the behaviour of the newly-hatched loggerhead turtle when it hurries from its cradle in the sand, where the mother buried the eggs, to its future home in the sea. It has an inborn preference for going down rather than up, and for blue rather than for any other colour, but careful experiment has shown that its chief life-saving impulse is to move away from the more blocked and interrupted horizon and towards that which is open and free. If the baby turtle is put in a tub, out of which it cannot see, it moves aimlessly; if it is put on a tub that has been turned upside down, it takes a look round and moves persistently towards the opener horizon. But it does not find the sea intelligently; it is *forced* to go right.

Very interesting are what may be called periodic promptings or rhythms that are somehow registered in the animal's constitution. On the flat beach at Roscoff in Brittany there is a minute green worm called Convoluta, which comes out of the sand in large numbers when the tide goes down, and disappears again at the first splash of the flow. When these tiny worms are transferred to a tideless aquarium they show the same rhythm as in the open; they continue appearing and disappearing with regularity for over a week. After that the rhythm wanes away, and long before that if the aquarium is darkened. In many cases, probably, even as high up the scale as the migratory birds, the bodies of animals have become attuned so that they keep time in some measure with the days and seasons and other periodic changes in the outside world. Thus one of the kinds of sea-urchin at Suez spawns at the full moon through the summer months. At this low level of life, where there are no nerve-centres, there can be no mental " memory " in our sense of the word, but there may be a *bodily* memory.

Following this line of behaviour we gradually reach " reflex actions," which sum up most of the behaviour of simple animals like sea-anemones and jelly-fishes. They correspond to what we do, " without willing it," when we draw our finger back from something very hot. Let us recall a previous study. Certain nerve-cells (sensory neurons) serve like "Scouts" to receive news from the outer world; they pass the thrill on to other nerve-cells (associative neurons), like " General Headquarters"; and from these by a third set of nerve-cells (motor neurons), like Executive Officers, the orders pass to the muscle-cells (effectors) which move. Thus the earthworm jerks itself back into its hole when the ground vibrates under the light tread of the blackbird's foot. Thus, again, the nestling opens its mouth at the touch of food in its mother's bill and then proceeds to swallow. No small part of animal behaviour consists of these reflex actions, differing from instincts in being usually over and done with in one movement, though reflex "A" may incite reflex "B" to activity and "B" may pull the trigger of " C." They are more automatic than instincts, and many of them do not require the help of the brain, some of the lower centres of the nerve cord or spinal cord being quite sufficient in themselves.

Instincts, obligatory movements, reflexes, and the engrained reactions of single-celled animals are on one line, all depending on something already arranged in the animal's structure. No doubt intelligent behaviour also implies pre-established structure, namely, more than a little in the way of a brain, but it means a power of adjusting the answers-back to suit particular circumstances, it means profiting by experience, putting two and two together; it means some experimenting and learning. There is a meeting of the two lines in some forms of what is called " trial and error " behaviour, when the animal tries its answers-back one after the other, seeking, as it were, the best solution. Thus the little trumpet-shaped Infusorian called Stentor which makes a protective case for itself on water-weed, was tested with a shower of microscopic dust. It bent to one side, but that was of no use ; it reversed the wafting action of its microscopic lashes or cilia, but the shower continued to fall ; it drew itself back into its sheath, but the experimenter was ready with more dust when it protruded itself again. Finally, it detached itself from its shelter and swam away. It had found a solution. But in some cases almost as simple, what the animal does is to try something new ; it makes an initiative or a fresh experiment, and this is a first step on that line that has its climax in intelligence.

Among common starfishes an individual sometimes tackles a small sea-urchin and disarms it gradually by wrenching off the snapping blades (pedicellariæ) that clinch on the soft suctorial tube-feet. When the sea-urchin is more or less disarmed, the starfish begins to protrude on it its very elastic stomach, which has poisonous as well as digestive juices. This is a particularly instructive case, for only some individuals among the starfishes will tackle sea-urchins ; and what is attempted must be persisted in, if it is to be of any use. But the starfish has no nerve-centres or ganglia, only a superficial nerve-strand up each arm, and a ring around the mouth connecting these strands and in some measure unifying their action, besides a network of nerve-cells in various parts of the body. Since there is no concentration of the nerve-cells into ganglia, we dare not use any big word like intelligence, we must keep to some term like *experimental.* But the starfish's behaviour in this case is on the same line as intelligence, and on a line different from that of instinct and the like, where the answer-back is not a fresh tentative but only part of an inborn repertory. It seems to us that these are the two main lines of animal behaviour.

If we hold an ostrich feather in our left hand sloping gently upwards, with the convex surface up and the concave surface down, with one set of barbs directed upwards and those of the other side downwards, we have a useful diagram of the diversity of animal behaviour. The upward-pointing barbs represent the various kinds of initiative, tentative, experimental behaviour, leading to a climax in the high intelligence of horse and dog. The down-turned set of barbs represent the various kinds of enregistered, engrained, reflex behaviour, leading to a climax in the marvellous instincts of ants and bees. There are good reasons for believing that there is a mental aspect through and through, and not merely in the higher animals where we cannot make sense of the behaviour without allowing for mind. The convex outer surface of our ostrich feather, as a whole and in each part, may typify the bodily or nervous activity that we can in some measure see; while the concave inner surface, as a whole and in each part, may typify the invisible mental or psychical life. But it must be noted that the mental aspect in animal life is not restricted to control of activities, it may express itself in feelings and pictorial purposings, and, among birds, in music and artistry.

CHAPTER XVIII

THE WEB OF LIFE

IN two previous studies we considered the bodily life of the individual animal and plant—how they keep agoing from day to day. This is the subject of the science of PHYSIOLOGY, which inquires into the " go " of the living creature. But this requires to be broadened by thinking of the bustle of life and all its intricate " give and take." This is often called ECOLOGY.

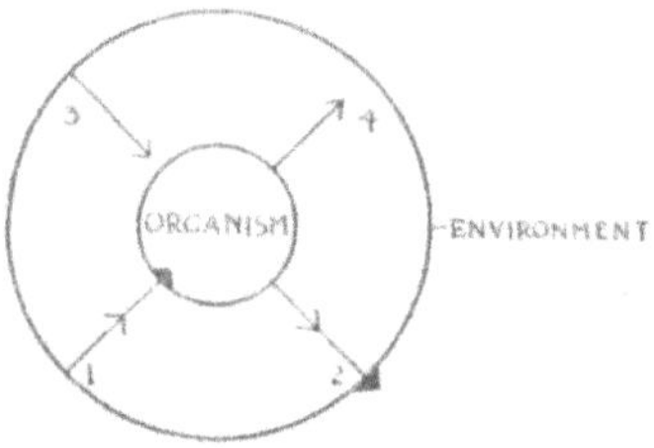

FIG. 21.

Diagram suggesting some of the relations between organism and environment. 1. An influence from surroundings may produce a modification on the organism. 2. An activity of the living creature may change the environment. 3. An environmental influence may affect the creature without any structural change. 4. An activity on the organism's part may affect the environment without causing any readily appreciable change.

Darwin's picture of the world of living creatures gives great prominence to the inter-linking of lives. Nothing lives or dies to itself. Every creature's life is like a circle intersecting other circles, and these intersections or inter-relations are often quite indispensable to the creature's daily life and to the continuance of its kind, while there are others, such as parasites and enemies, which the animal would be glad to get rid of, if it could. Let us think over some examples of what Darwin meant by the web of life.

Living always implies a give-and-take between the creature and its surroundings, or, to put it more technically, between the organism and its environment. Indeed, we might call this the unending problem of life,—to establish and keep up good relations between the living creature and its surroundings. But the surroundings of an animal include other animals, and in most cases plants as well; and the surroundings of a plant include other plants and also animals. Thus arise the countless intersections or inter-relations which we are discussing.

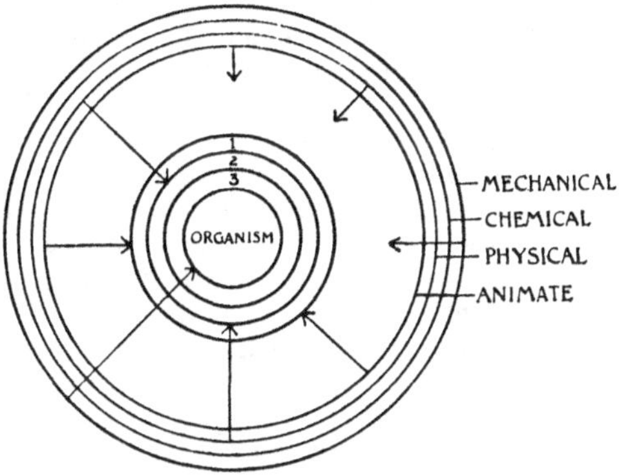

FIG. 22.

Diagram suggestive of the influences of the environment that play upon living creatures. The environment is represented by four circles—mechanical, chemical, physical and animate. The organism is represented by three circles, (1) the ectoderm, e.g. the outer skin and the nervous system, (2) the mesoderm, e.g. the muscular system, and (3) the endoderm, e.g. the food-canal. Some influences operate mainly on (1), some mainly on (2), others on (3).

One of the commonest of intersections is, that one living creature uses others as food; and this leads to what may be called nutritive chains. The robber or Skua gull chivies the herring gull and induces it to part with the mackerel that it has caught and swallowed; the mackerel feeds daintily on minute crustaceans called Copepods; these depend on microscopic animals, such as the Infusorians called Peridinids, and on microscopic marine plants, such as the beautiful cells called Diatoms. Thus there is the nutritive chain: Skua Gull, Herring Gull, Mackerel, Copepods, Infusorians, and Diatoms.

Living Nature is run, so to speak, on a plan of unending reincarnations. Dead vegetation in a box by the side of a pond may be broken down or rotted by bacteria which batten upon the decay. There is no rotting without bacteria. When they overflow in a living cascade into the pond, they form food for the simple animals called Infusorians. But these in turn are eaten by minute crustaceans or "water-fleas," which in their turn form sustenance for fishes. As the fishes may serve as an excellent brain-food for man, we see a long chain from decaying vegetation to the part of man's body in which clear thinking is at home. What was part and parcel of the decaying bracken thrown into the pond passes through the invisible bacteria to form the substance of just visible Infusorians; these are re-embodied in pinhead-like Copepods; these find re-incarnation in Fishes; the last avatar is in Man. If the fisherman should lose the contents of his basket into the water, other bacteria would reduce the dead fishes into gases, salts, and water, which might again form the food of plants. Thus a new cycle commences. Truly no organism either lives or dies to itself. As the old Greek philosopher, Heracleitus, said: "All things flow."

Next to the nutritive chains that bind one animal to others, which it eats, or which try to eat it, there are the linkages which secure the continuance of the race. Thus the majority of flowering plants depend, as we have already noticed, on the visits of insects for bringing in the fertilising dust or pollen which makes a possible seed or ovule into a real seed or embryo plant. Inside the ovule in the seedbox of the flowering plant there is an egg-cell, and this will begin to divide and develop if it be fertilised by a male element which is carried by the pollen grain. It is therefore in most cases necessary that an insect visitor, such as a bee or butterfly, should dust the stigma of the flower with pollen collected from the stamens of another flower of the same kind. The collecting of pollen is often quite unawares, the hairs of the body of the insect entangle some of the minute grains; the dusting of the stigma is almost always quite unawares, though there are a few cases, like the Yucca moth, where the visitor behaves very methodically if not deliberately. The linkage between flower and insect is the most important linkage in the world, for it not only secures the continuance of the race of flowering plants, but also secures cross-fertilisation, which means a greater yield and a stronger set of seeds than usually results if a normally cross-pollinated flower is artificially pollinated with its own pollen. The flowers and their visitors are adapted to one another as glove to hand. It should be noticed, however, that some flowers, like those of grasses, are fertilised by wind-carried pollen, that a few, like the pea, fertilise themselves, and that a very few, like the dandelion, do not require to be fertilised at all.

The race of fresh-water mussels cannot continue unless the pinhead-like bivalved young ones, that are set free from the mother mollusc, are able to attach themselves for a while to the skin of a fresh-water fish, such as the minnow. And, strange to say, there is a continental fresh-water fish, called the Bitterling (Rhodeus amarus), which inserts its eggs into the gill-plate of the fresh-water mussel, where they pass through the early stages of their development. Similarly, the microscopic young stages of the liver-fluke, an often serious parasite of sheep, require to sojourn for a time inside a fresh-water snail; and in Britain there is only one species, Limnæa truncatula, that serves this purpose. Since water-wagtails and lapwings are fond of picking up and swallowing the little water-snail, we can at once see the connection between birds and sheep-farming. Little fishes are fond of the aquatic larvæ of gnats and mosquitoes, and they are often introduced into water-reservoirs in India and other warm countries, for they devour the mosquito larvæ, and thus reduce the number of adult mosquitoes which are responsible for carrying the microscopic malaria organism (Plasmodium) from an infected patient to a new victim. Thus there is a very clear linkage between little fishes and human welfare.

Another common intersection of life-circles is seen when one living creature serves to scatter another. In our study of autumn we have noticed that the seeds of many juicy fruits are distributed unwittingly by birds, for while the pulp of the fruit is digested in the food-canal, the uninjured seeds are passed out and may thus be widely sown. Minute water-plants and water-animals are often carried on the wet feet of birds from one lake to another. From one clodlet on a red partridge's foot Darwin got eighty seeds to germinate. Prickly fruits like those of Jack-run-the-hedge, with its other appropriate name of "Cleavers," become attached to passing animals and are carried from place to place, until they eventually fall off.

It often happens that two organisms become very closely wrapped up together in the bundle of life. Thus the heather succeeds so well on the poor soil of the moorland, because it has entered into partnership with a fungus that finds its way through the whole body from root to flower, and serves as a middle-man between the green plant and the sour soil-water. All the plants of the pea and clover tribe have partner bacteria that form galls or tubercles on the roots, and somehow enable the plant to capture the free nitrogen of the air entangled in the soil. The young death-watch beetles that bore in old wood are able to thrive on this dry-as-dust diet because they have crowds of fermenting partner-yeasts in their food-canal.

But perhaps the meaning of the web of life may be best illustrated by taking a particular kind of animal, and we select a very interesting case.

White ants or termites are seldom very white and are certainly not ants, but they are perhaps the most interesting insects in the world, which is saying a great deal. They are creatures of long pedigree, in an order by themselves, distantly related to the ancestors of the cockroaches. They have an extraordinary social system, of which they are probably but dimly aware, for their brains are relatively small and their behaviour is almost entirely instinctive, requiring no apprenticeship and very rarely showing any gleam of intelligence. Besides the "king" or male and the "queen" or female—the parents of a community which often includes half a million—there are two kinds of reserve or complemental royalties, who may replace the king and queen if these should die. Then there are the normally sterile workers and soldiers, arrested individuals of both sexes, unlike the workers among ants, bees, and wasps, which are arrested females. There may be two distinct sizes of worker and three distinct sizes of soldier! In some cases there is a quaint type of soldier without the usual big jaws, but able to squirt a glue-like secretion in the face of assailants, such as true ants. The big-jawed soldiers cannot chew wood as the workers do, so they have to be fed by the workers; and it may be noted that there is much give-and-take in the community, for the members of one caste often feed those of another, and often lick one another for the sake of secretions that exude on the thin-skinned body.

What concerns us here, however, is not the domestic economy of the termites, but the way in which the circle of their life intersects other circles. The great majority feed on decaying wood and vegetation, and thus they assist in the never-ceasing circulation of matter. They could not thrive on the dry-as-dust food afforded by decaying branches unless they had, in their food-canal, numerous partner Infusorians, which prepare the material for use. By raising the temperature experimentally it is easy to kill the Infusorians without harming the termites, and thus

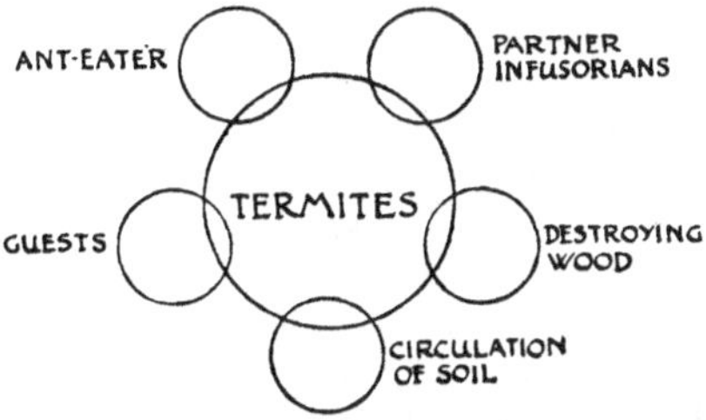

FIG. 23.

A universal ecological diagram showing how the circle of one life, e.g. that of Termites, may intersect the circles of many other lives.

it has been demonstrated that the presence of the partners is absolutely indispensable.

Most worker termites are more or less blind, the eyes being absent or reduced to little points ; and all save a few exceptional forms are perturbed by the glare of day. Almost all of them like darkness, slight humidity and stagnant air—conditions which are fulfilled by the usually substantial termitaries, built of chewed earth, or wood, or both. A ground-nest may be ten feet high, and strong enough to bear a man's weight, but sooner or later there is weathering, and thus the soil is kept in circulation. The same is true of the often very hard tunnels of chewed earth which the little blind workers construct up the stems and along the branches of trees. In the course of time the particles give way and are swept by the torrential rains to add to the fine soil of a distant valley. In short, termites have some share in soil-making, though their importance in this connection is not nearly so great as that of earthworms.

Although white ants are not related to true ants, except that both are insects, there are very remarkable parallelisms in habit. Thus we find in both groups an occasional cultivation of edible fungi, and some of the termites make intricate labyrinths of chewed wood on the walls of which the useful fungus is grown. Various kinds of insects, called "termitophiles," live in association with termites, just as others do with true ants; and in both cases the habitual associates may be classified as injurious intruders, parasites, tolerated guests, and esteemed pets. The last exude secretions which are greedily licked off by the termites. In exchange the pets sometimes receive food, but there are some that help themselves to young termites, of which there are plenty and to spare. A queen termite may have ten million offspring in a year and live for ten years.

Most of the insects cherished by termites are small beetles, but there are representatives of some other orders, such as flies. Among some of the guests an extraordinary half-diseased condition appears. It involves a swelling of the posterior body, an accumulation of fat, blindness, and winglessness, and is regarded by Professor Wheeler as a direct consequence of the cramped quarters, the absence of light, the stuffy air, and the abundance of carbohydrate food. It is the seamy side of enjoying hospitality.

Termites are devoured by many insect-eating mammals, birds, and reptiles; and true ants are their inveterate enemies. So they are always passing into a new avatar. Their largest enemy is the Cape ant-eater or Aard Vark which breaks into the termitaries by night and whips in thousands of workers on its long worm-like sticky tongue. On the other hand the termitary may afford more or less casual shelter to scorpions, snakes, lizards, and even birds, but the details of the inter-relations are unknown. One would expect in most cases a speedy change of lodging, for the termites' jaws are irresistible, and an irritant and corrosive secretion exudes from the mouth.

If we could imaginatively insert one leg of a giant pair of compasses in a termitary and describe a circle on a big scale with the other, what a variety of life-circles we should intersect: trees and shrubs, fungi and Infusorians, ants and ant-eaters, beetles and birds, and so on and on. The circle of human life is of course intersected at many points. The termites devour floors and rafters and furniture, boxes, books, and papers. It is futile to erect wooden telegraph posts, and we have heard that there are places where it is dangerous for a man with a wooden leg to rest too long. The Australian bushmen make temporary ovens of the termite mounds and may even eat some of the salivated clay. The Hillmen of India eat the termites themselves, and the queen with her four-inch long abdomen, swollen with eggs and food, is said to be delicious. The earth of the termitary is often ground to powder and used as a basis for tennis-courts and the like. How Darwin would have enjoyed the modern disclosure of the intricate inter-relations between termites and other living creatures, for it is but an extension of one of his central ideas—the inter-linking of lives in a web of life.

If man is to control Nature and preserve its balance, he must carefully respect the web of life, both when he introduces new plants and animals into a country, and when he destroys. How calamitous has been his introduction of rabbits into Australia and sparrows into the United States. How far-reaching, often, have been the costly consequences of destroying useful birds, that keep down injurious insects, and useful mammals, like weasels, that check the multiplication of mice and voles. Theoretically the fact of the web of life is very important, for there is continual sifting and winnowing in Wild Nature, and this necessarily works in reference to the already woven web of life. This makes the sifting very subtle; it determines, so to speak, the survival of animals that can say Shibboleth and the elimination of those that can only say Sibboleth!

CHAPTER XIX

THE CURVE OF LIFE

OUR life has sometimes been compared to an arched bridge, rising to a short level middle portion and then descending again. More simply, it is comparable to an ascending and descending curve,—a trajectory. For there is a time of developing, growing, strengthening, mellowing ; and then comes the time of decline, failing,

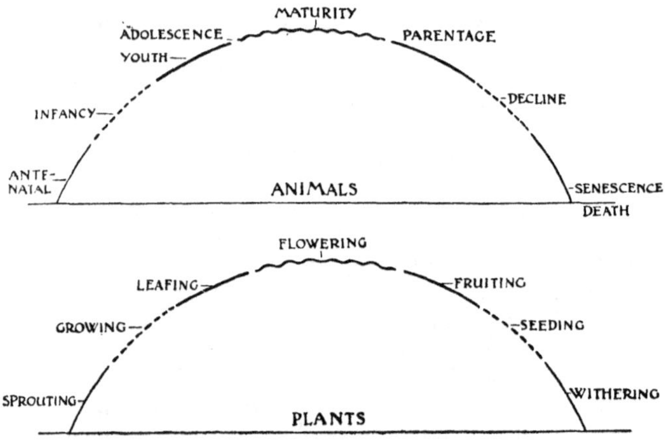

FIG. 24.—COMPARISON OF THE CURVE OR TRAJECTORY OF LIFE IN PLANTS AND ANIMALS.

ageing, and dying. This must be more or less true of all living creatures, but the details of the curve differ in different types, and there are minor ups and downs on each successive arc.

The animal begins usually as a fertilised egg-cell, containing in some way that we cannot picture the whole inheritance. This egg-cell undergoes cleavage or segmentation, and develops into an embryo, the cells showing a remarkable division of labour. Some become nervous and others contractile, some digestive, and others skeletal, and so forth. But some cells do not share in the body-making, and are set apart, sometimes very early, to become the reproductive cells of the animal that is being developed.

Eventually the developing organism or *embryo* is hatched, bursting out of the egg-shell or egg-membrane, while in other cases, where it is hatched within its mother, it is set free or " born " as a young creature. The word embryo is used for the developing organism before it is hatched, or before it is born. A plant's seed is an embryo, comparable to the unhatched chick.

What comes out of a hen's egg is a fully-formed and adventurous chick, which we must simply call *the young bird;* but what comes out of a butterfly's egg is a caterpillar, and to this we apply the general term *larva.* Tadpoles, grubs, caddis-worms, newly hatched crabs may be mentioned as good examples of larvæ ; and the peculiarity is that their structure is very different from that of their parents. Or, to put the same thing in another way, the larvæ cannot develop into adults without some *metamorphosis.* In most cases these larvæ, which complicate the life-history, are to be regarded as interpolated stages that are specially adapted to some advantageous end. Thus they may be able to shift the young animal to safer surroundings, as when the newly hatched shore-crabs swim out into open waters, away from the rough-and-tumble struggle of the shallows. Or they may be suited for voracious eating, as in caterpillars, and for the accumulation of reserves that make a vigorous adult life more attainable. Thus the aerial winged life of many insects is entirely devoted to " love," while the larval life is entirely devoted to satisfying " hunger." In other cases the larva is in part an illustration of the general tendency of individual development to recapitulate racial history. Thus the young larvæ of flat-fishes, such as soles, are quite symmetrical little creatures just like the larvæ of haddock and herring ; and there can be no doubt that the flat-fishes are descended from ancestors of the ordinary fish form. Similarly, after a short free-swimming phase, the larva of the rosy featherstar or Antedon becomes fixed and stalked just like a miniature of the sea-lilies from which the free-moving adult Antedons are derived.

In many animals the delicate early stages, comparable to human infancy, are succeeded by more robust youth, comparable to human childhood. During this period in some cases there is play, important as an irresponsible apprenticeship to the business of life, as in kittens with their sham hunting, and also as affording elbow-room for the expression of novelties and originalities. Play is likewise a safety-valve for overflowing motor energies, and an opportunity for learning the lessons of "give and take" and self-subordination in a team—advantages which are important for many an animal as well as for mankind.

By and by comes the period of growing-up or adolescence when childish characters are put off and adult characters are put on. There are subtle re-arrangements in the body, especially in the nervous system, and sex begins to have more influence on life. Following the curve a little farther, in birds for instance, we come to falling in love, or, as it should be called, *rising* in love—a great chapter in the trajectory. The organism gains its full strength and in natural conditions it becomes a parent. This is the top of the arch on the life-bridge, the crest of the ascending and descending curve.

Then comes ageing, sometimes quickly, sometimes slowly. "And so from hour to hour we ripe and ripe, and then, from hour to hour we rot and rot, and thereby hangs a tale." The single-celled simplest animals or Protozoa seem to be able to evade ageing and natural death; but all ordinary animals grow gradually old. All through life there is a contest going on between waste and repair, work and rest, running down and winding up, senescence and rejuvenescence. Gradually, however, there accumulate fatigue-results of wear and tear that are not made quite good by food and sleep, rest and change. Slowly the tortoise creeps up on the hare, and ageing begins to gain on keeping young. Thus the curve inevitably sinks towards death.

Let us picture, then, a common animal life-curve—embryonic development, tender youth, larval period, "childhood," adolescence, love-making, maturity, parentage, full strength, ageing, senescence, death. Parallel to that may be pictured the normal curve for higher plants:

sprouting, growing, leafing, flowering, fruiting, seeding, withering, and dying.

It is a very interesting exercise to try to construct a life-curve for different kinds of animals, for they differ from one another in the proportionate length of the successive arcs. One animal may have a prolonged youth and a very short maturity, while another has a hurried youth and a very leisurely ageing. In the case of may-flies there is a prolonged youth of two, three, or sometimes four years, then perhaps one evening or a couple of evenings of active adult aerial life, reproducing without feeding, "love" without "hunger." Then comes an abrupt fall to death. There is one species of may-fly which has actually but a single hour of adult life—a long larval period but the whole of the full-grown life telescoped into one brief hour. This may-fly or day-fly is not even an ephemerid!

But contrast with the may-flies the mound-birds of Borneo, where youth has been telescoped. The egg is laid in a great heap of fermenting vegetation, where it hatches out without any brooding. When the young bird scrambles out of the egg-shell, it instinctively continues its struggles and frees itself from the mound. An hour after hatching, it is fending for itself in the scrub! It has passed out of the egg into maturity in one great bound, skipping the whole juvenile period.

In contrast to this, again, think of a highly endowed mammal like the elephant, which sometimes remains beside its mother for ten years, a long period without great responsibility. But the prolonged youth is in this case followed by a prolonged maturity. Or recall the strange case of the common eel. The young eels or elvers which come into the European rivers from distant Atlantic waters are already two and a half or two and three-quarters years old. In the course of their long journey from a stretch of sea not far from the Bermudas, they have undergone a metamorphosis from knife-blade-like glass-eels (called Leptocephali) to cylindrical elvers. These ascend the rivers and continue to develop and grow in the quiet waters. They become full-grown and ripe in five to eight years, and then make for the open sea—a second long journey—apparently ending in sudden death after spawning. This is another case where the infantile and adolescent periods are greatly prolonged, but the adult life is relatively short, and stops abruptly. It is a good exercise to picture to ourselves a number of these life-histories or life-curves.

But now, with the picture of the life-curve clearly in our mind, let us go back again to the beginning of the individual life. Almost every many-celled animal, from sponge to man, begins its individual life as a fertilised egg-cell. Only a few exceptions have to be allowed for, as when a fresh-water Hydra multiplies by budding, or a simple worm divides into two, or a drone-bee develops from an egg-cell that was not fertilised by a sperm-cell. For while the drone has a mother, it has no father ; and the same is true of the summer green-flies or Aphids that are so common on bean-plants and rose-bushes.

Similarly in the plant-world, the oak-tree may be traced back to a sapling, and the sapling to a seedling, and the seedling to a seed. But the seed is an embryo-plant that develops from an egg-cell that lies in the recesses of the possible seed or ovule. Just as with animals, this egg-cell has to be fertilised by a male element, which, in this case, is a nucleus that comes from a pollen-grain dusted on to the pistil of the flower. In rare instances, such as dandelions, the plant egg-cell can develop without being fertilised, just as in the case of the summer green-flies. This is called parthenogenesis.

It should be noticed, however, that in many flowerless plants, such as ferns and mosses, there are two clearly-marked generations, one developing from an unfertilised spore-cell and the other developing from a fertilised egg-cell. In the same way certain animals illustrate a similar *alternation of generations.* Thus a fixed rather plant-like zoophyte buds off swimming-bells or medusoids, whose fertilised egg-cells develop into zoophytes again.

In some way that we cannot picture, all the hereditary characters of a living creature lie latent in the fertilised egg-cell. We cannot compare this to anything else, for it is quite unique. We might think of a bud, in which there lies the possibility of a shoot and leaves, or of a flower. But if we open a bud we find the visible young beginnings of the parts that will be unfolded, and this is what we do *not* find in a fertilised egg-cell. Later on we may find the embryo chick inside the egg-shell, but to begin with there is nothing but a clear drop of living matter lying on the top of the yellow yolk. In that clear drop there lies the whole inheritance, and out of that clear drop the chick is developed. So development means that out of the apparently simple there comes the obviously complex. It means making the invisible visible, the latent patent; but it is not really like anything else in the world.

There must be in the living matter of the egg-cell, produced by the female, and of its counterpart, the sperm-cell, produced by the male, something in the way of initiatives which are able to give rise to the characteristic features of the developing organism; and it is usual to call these initiatives "hereditary factors" or "genes." As we noticed in the Tenth study on "the beginning and basis of life," these little hereditary "somethings" are carried, in whole or in part, by the nuclear rods or chromosomes. It is an extraordinary fact, difficult to believe, that in a few much-studied cases, such as the fruit-fly, it is now possible to map out the arrangement of the hereditary factors as they lie in absolute invisibility on the chromosomes, like beads on a string!

Each kind of living creature has a particular number of chromosomes in each cell of its body. Thus the smallest number, two, occurs in the threadworm parasite in the horse, while the number for the horse itself is sixty-four, and for man forty-eight. But in the ripe egg-cell and the ripe sperm-cell the number is always half the normal, for there is a process of halving in the course of the ripening. If the unripe egg-cell has eight chromosomes, the ripe egg-cell has four; and similarly in the case of the sperm-cell. Therefore when the two come together and unite in fertilisation the normal number is restored. As each ripe germ-cell contains a complete or nearly complete equipment of hereditary factors, the combination of maternal and paternal contributions means that the fertilised egg-cell has a duplicate inheritance, half from the mother and half from the father. What Huxley prophesied in 1878 has now been proved up to the hilt, that the developing organism is like a web, with the warp from the female and the woof from the male. But it is not by any means the case that the maternal and paternal contributions need find equal expression in the course of development.

When the characters of the two parents are both expressed and, as we might say, averaged, the inheritance is said to be *blended* or *apparently blended*. Thus Cheviot sheep crossed with Leicesters yield very useful sheep called Half-breds, which show characters of both, and breed true. When a rabbit with long lop ears, reaching to the ground, is crossed with a rabbit whose lop ears are short, the offspring have lop ears whose length is an average of that shown by the two parents. A mulatto seems to be in various ways a blend between the white father and the negro mother, but this is not such a clear case as it seems at first sight.

In other cases the pooling of paternal and maternal characters produces in the offspring a very coarse-grained combination, which is called *particulate.* The characters of the father appear in one part of the offspring's body and those of the mother in another part of the body. This is sometimes very clearly seen in a piebald pony. Quaint instances sometimes occur in sheep-dogs, where the mother's eye looks out of one side of the offspring's head, and the father's out of the other!

Yet another mode of inheritance recognised by Mendel in 1865, overlooked till 1900 and then re-discovered, is of great practical and theoretical importance. It is called *Mendelian* inheritance, and it is illustrated by well-marked crisply defined, non-blending features which are called "unit characters." Thus the Hapsburg lip which persisted for centuries in the royal houses of Austria and Spain is a well-known instance of a unit character. Some people have fingers all thumbs, and this peculiarity ("brachydactyly") is a unit character, which has been known to recur in a certain percentage of a lineage for six generations. Such a subtle peculiarity as "night-blindness" or inability to see in dim light has been traced through a family history back to one Jean Nougaret who lived when Charles the First was king. The difference between wood-snails with banded shells and those with bandless shells depends on a unit character, and similarly with giant and dwarf garden peas, crested and non-crested poultry, normal and albino guinea-pigs, ordinary and waltzing mice. Hundreds of these unit characters have been studied; they are crisply defined, they do not blend or break up, they are either there or not there in the offspring, they illustrate Mendelian inheritance. As the recognition of this kind of inheritance has illumined the whole subject of heredity, and is one of the big discoveries which this book is meant to illustrate, we must try to understand in a general way what Mendel discovered.

Suppose we have to deal with two animals of the same species that differ from one another in some clear-cut unit character, of the non-blending type. Let us say that one is an ordinary mouse and the other a dancing mouse, which has some curious peculiarity in its nervous system. Thus on the slightest provocation or none it waltzes round in a circle. If the two animals are paired, and if there is a family, the offspring will all appear quite normal; and this fact is technically stated when we say that normalness turns out to be *dominant* and waltzingness *recessive*. But we cannot tell beforehand whether a particular character will be dominant (i.e. expressed) or recessive (i.e. suppressed). This is step one.

If the offspring that result from the cross—called the first filial generation, F_1—are paired together, or with others of similar history, then in the next—the second-filial, F_2, generation—25 per cent. of the offspring will be pure waltzers, 25 per cent. will be pure normals, and 50 per cent. will look like pure normals, but will be like their immediate parents in their constitutional make-up. We mean by a "pure waltzer," that if this mouse is paired with others like itself, and that continues, the lineage will never include anything but pure waltzers. Similarly with the "pure normals." But if the 50 per cent. *apparent* normals, technically called "impure dominants," be paired with others like themselves, they will show in their progeny the same proportion of types: 25 per cent. pure normals, 50 per cent. apparent normals, and 25 per cent. pure waltzers.

This Mendelian rule may be stated in a simple scheme:—

(Parents) . D × R . (D with the dominant, R with the recessive character)

(First Filial Generation, F_1) D(R) . (the R character latent or unexpressed)

× × i.e. (pairing with one another)

(Second Filial Generation, F_2) × 25DD + 50D(R) + 25RR ×

F_3 . . . DD 25DD+50D(R)+25RR RR

As instances of dominant and recessive Mendelian characters we may mention the following, putting the dominant always first: hornlessness and hornedness in cattle, short hair and angora hair in rabbits, crest and absence of crest in poultry, greyness and whiteness in mice,

FIG. 25.—MENDELIAN INHERITANCE IN ANDALUSIAN FOWLS.

A black and a white are crossed; the offspring are "blues." If "blues" are bred together there will be on an average in a dozen offspring, three blacks, three whites, and six "blues."

pink eye and white eye in fruit-flies, tallness and dwarfness in peas, yellow seed-leaves and green seed-leaves in peas, absence and presence of awn in wheat, susceptibility and immunity to rust-disease in wheat, markedly dentate and slightly dentate leaves in nettles.

Let us take another instance in more detail. If black Andalusian and white Andalusian fowls be bred together, the offspring (F_1) bear a finely divided pattern of black-and-white markings which give a blue effect (" imperfect dominance "). If two of these blue Andalusians be mated, their offspring (F_2) show in an average dozen three pure black, three pure white, and six " blue." In this case the " impure dominants " are externally distinguishable from the pure dominants, the blacks.

It should be noticed that Mendel did much more than state this rule of inheritance, he explained how it came about; but that is beyond our scope in these studies. What we have tried to do is simply to indicate how the curve of life is becoming more and more intelligible. What a living creature becomes is dependent on its hereditary nature and also on the nurture that this receives. By nurture, which has become a half-technical word, is meant all the influences of surroundings, food, and habits. If we think again of the bud that is going to become a flower, we may say that the success of the flower is a product of the nature and the nurture. Much depends on the original vigour of the bud, but the unfolding of the blossom, which may be rich or poor, depends partly on the nurture, such as the sunshine and the rain. What the organism becomes is always the resultant of two components: the inherited " nature " and the available " nurture."

CHAPTER XX

MAN IN THE MAKING

IN previous studies we have asked how the earth came into existence, and how plants and animals evolved and spread into every corner of land and sea. We must now ask an even more difficult question: How did Man arise? The answer is part of the science of ANTHROPOLOGY.

When we set ourselves to think of Man's origin and pedigree we should think first of his uniqueness. In the familiar words of Shakespeare's appreciation: "What a piece of work is a man! How noble in reason! How infinite in faculty! In form and moving how express and admirable! In action how like an angel! In apprehension how like a god!" And yet we cannot always recognise ourselves in this fine dress, which often expresses what we wish to be rather than what we are. Many uncivilised peoples are very dull, though seldom so dull as the careless traveller thinks. Sometimes it is the traveller who is dull! Moreover, we know that before the modern man type, called Homo sapiens, appeared on the scene there were slouching men of Neanderthal, Homo neanderthalensis, short of stature and slow of speech, but big-headed and with beetling brows, who used fire and buried their dead with reverence. According to most authorities these Neanderthal men, so called because of the place, near Düsseldorf in Germany, where some of their remains were found, were no ancestors of ours, but distant cousins, who dwindled away without leaving any descendants. They were true men, of the genus Homo, but not of our species. In any case they point backwards to still humbler forerunners, who could not be called more than half-men or "tentative men."

Modern man can weigh the heavens in a balance and measure not only the suns and stars, but the circumference of the universe. He can knock a fragment out of an atom, and conjure with the powers that issue from radio-active substances like water welling from a spring. He plays upon the long gamut of electro-magnetic waves, discussed in another study, and uses the longest for broadcasting and the shortest for healing his sick. Since he makes soil-fertilisers by sending terrific electric discharges through the atmosphere, he may be said to wring bread out of the thin air. He is almost like a creator in building up very complex substances, like indigo or thyroxin, out of very simple materials. Among plants and animals he begins to control the generations yet unborn; he has turned a wolf into a dog and a big-seeded mountain grass into the wheat of a thousand fields. For himself he has begun to control the future with deliberation, and he has conquered many of the diseases that his flesh is heir to. We are bound to go entirely wrong in our thinking if we do not picture man as great in performance and in promise, as standing in some remarkable way apart from the rest of creation. And yet how can we draw any line between Man as conqueror of his kingdom and the primitive men, our ancestors of long ago, of whom Æschylus drew, in "Prometheus Bound," such a vivid picture: how "first beholding they beheld in vain, and, hearing, heard not, but, like shapes in dreams, mixed all things widely down the tedious time, nor knew to build a house against the sun with wicketed sides, nor any woodwork new, but lived like silly ants, beneath the ground, in hollow caves unsunned. There came to them no steadfast sign of winter, nor of spring flower-perfumed, nor of summer full of fruit, but blindly and lawlessly they did all things." And when we work back to primitive men, we cannot stop there, but must continue gropingly to men in the making, the Hominoids, as we may perhaps call them; and beyond these, in thicker mist, we must feel our way cautiously to an ancestral stock common to the Hominoids and the Anthropoid Apes.

What we are surest of is man's greatness, and the true inwardness of any living creature is always to be looked for in the highest level, not the lowest level, that is reached. What is Man? is a question which we must not think of answering before we have added to a study of savages, our "contemporary ancestors," and of primitive man, who is but dimly descried, an appreciation of man at his best, in

Plato and Aristotle, in Newton and Darwin, in Shakespeare and Goethe. We must think of man as the maker of Science, the discoverer of Nature's secrets, as an artist who gives shape and sound to ideas which scientific descriptions cannot express, as a man of goodwill, who, at his best, is always searching after the true, the beautiful, and the good, and enlisting comrades in his quest. Only when we think of what man may be is it safe for us to look long at the rock whence we were hewn and the pit whence we were digged. We must not blink any facts, however, and the scientific conclusion clearly indicated is that men in the making split off very long ago in the Miocene period—probably not less than one million years ago—from a stock common to them and the large apes of to-day. No one believes that tentative men sprang from any *existing* apes or monkeys. That is hopelessly absurd. The scientific conclusion is that men and apes diverged from an extinct common stock of which little is known; the Anthropoid Apes went one way—magnificent creatures some of them are—while the Hominoids went another way, which found its climax in Homo sapiens.

What evidence is there that man is distantly related to Anthropoid Apes? There is a striking sameness of structure between man and the highest apes, bone for bone, muscle for muscle, brain-wrinkle for brain-wrinkle. There is no important part of our brain that is peculiar to us, though many parts of the human brain, compared with the same parts in a gorilla's brain, are much larger and more complex.

Moreover, we are walking museums of relics, some of which link us back to ape-like or Simian ancestry, while others point to antecedent forms much more remote. The dwindling tag or "third eyelid," at the inner corner of our eye, and the rarely movable muscles of our ear-trumpet are familiar instances of our numerous vestiges, eloquent of the past, which still lives, or echoes at least, in the present. These vestigial structures are like the unsounded letters in many words, like the "o" in leopard and the "b" in doubt. These letters are functionless, but they give us a hint as to the history of the word. Or the vestigial structures in our living body might be justly compared to vestigial items in our clothing, like buttons without holes and holes without buttons,—useless now and dwindling, for the button-hole is often permanently closed. These vestiges are tell-tale relics of the distant past when they were well-developed and of everyday service.

There is a striking sameness in the bodily life of man and ape, though their mental life is normally so different, and they sometimes come near one another in their misery, when they are victims of the same disease, such as tuberculosis or rheumatism. Moreover, human blood may be transfused harmoniously into a chimpanzee, though not into a monkey. The early development of man is very like that of an ape. Hark-backs or reversions, and cases where the human structure is not fully finished, often betray Man's distant relationship with the Apes.

With facts like these before us, we must use the words with which Darwin concluded his "Descent of Man": "We must, however, acknowledge, as it seems to me, that man, with all his noble qualities, with sympathy that feels for the most debased, with benevolence which extends not only to other men, but to the humblest living creature, with his God-like intellect, which has penetrated into the movements and constitution of the solar system—with all these exalted powers, man still bears in his bodily frame the indelible stamp of his lowly origin."

Darwin spoke frankly of "the great break in the organic chain between Man and his nearest allies," but this gap does not seem quite so great to-day. The gulf between Homo sapiens and the earliest men has been bridged, so to speak, by the discovery of remains of more primitive human types —the Neanderthal man, the Rhodesian man, the man of Taungs, and the Heidelberg man. The gulf between Homo and a vaguely known stock common to the Anthropoid Apes and to Homo has been spanned by the discovery of tentative men, notably Eoanthropus from Piltdown in the Sussex Weald and the Apeman from Java, Pithecanthropus the Erect.

There is a solemnity in the patience of the age-long man-ward adventure which has crowned the evolutionary process upon the earth—though we must never think of evolution not going on. Perhaps three millions of years ago, probably from among Insectivores like the quaint Tree-Shrews and the Spectral Tarsier, there evolved the half-monkeys or Lemurs, the beginnings of the monkeyish or Primate order in which the brain became more for vision and less for smell than ever before ; and many acquisitions were made in connection with life in the trees. The fore-limb was liberated from the task of being a support and lever for the body ; it became the bearer of a free hand—reaching out, probing, testing, grasping. There was a marked re-

FIG. 26.—THE SPECTRAL TARSIER, *TARSIUS SPECTRUM.*

A unique large-eyed Malayan animal, arboreal and nocturnal, often ranked as a Lemur. It represents one of the steps on the path of brain evolution that has reached a climax in man. The fingers and toes end in peculiar expanded discs.

duction of the snout, leaving room in the skull for a larger brain, and the forward-shunted eyes, looking no longer sideways, made a more stereoscopic vision possible. The region of the brain set apart for skilled manipulation, visualising, attention, and still higher faculties began to be much more strongly developed.

From amongst many facts let us single out this one. If we compare man's brain with the gorilla's, the three most marked differences in size concern the areas that have to do with the control of precise muscular movements, with speech, and with the understanding of speech. Now these are the three areas which are the last to attain their full development in the unborn child.

Between three and two million years ago the Primate stem sent out its first tentative branches, such as the squirrel-like Marmosets; and the result was what might almost be called a tangle of monkeys. First the New World monkeys diverged, and then those of the Old World; the main stem, still very shadowy to us, for there are so few fossils, *grew on*. It next gave off the branch of the smaller Anthropoid Apes, the Gibbon and the Siamang, and then that of the higher Anthropoid Apes, Chimpanzee, Orang, and Gorilla. The main stem, vaguely known, though there are some fossils, *grew on*. Without haste, without rest, and sometimes leading to nothing, more branches were given off—and eventually there emerged the "tentative men" Hominoids, but not Homines, such as Pithecanthropus the Erect in Java and Eoanthropus in the Sussex Weald. At last came Homo, but even among his species there was the same process of sifting, for several, like the Neanderthal men, who shared in the struggle, failed to enter into the promises. When we envisage the long sifting-out process, we feel the vulgarity of saying that "man sprang from a monkey."

The study of evolution is still very young, and it is too much to expect evolutionists to be ready to answer the question: How did Man emerge or arise from an ancestral stock, common to him and to the higher apes? About a million years ago, Hominoids or "tentative men" began to go one way, while Anthropoids, like the ancestors of the chimpanzee and gorilla, went another way. What made the difference? But we must not be in a hurry to answer. We can only submit certain facts to be thought over.

First of all, there is the fact that in the Miocene Period, when the forking of the ways occurred, there was a notable increase in brain capacity in a number of unrelated types of mammal, elephants for instance. The reason is unknown, the fact is certain.

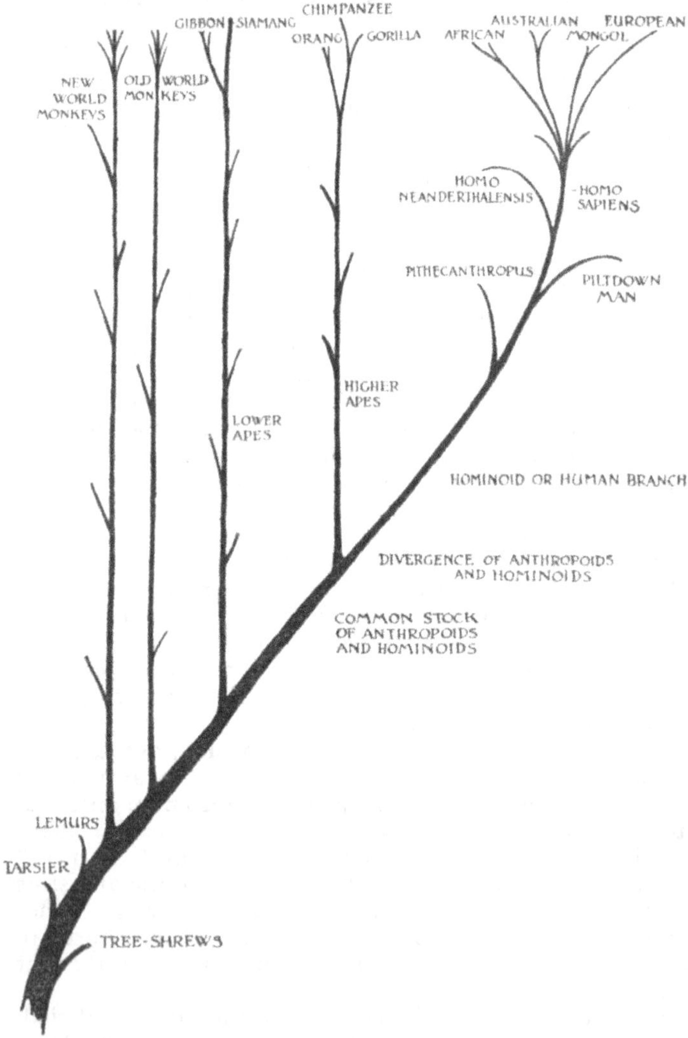

FIG. 27.

Diagrammatic indication of man's genealogical tree, based on the work of Sir Arthur Keith.

Second, there is evidence of a particular trend of brain-evolution, worked out by Professor Elliot Smith, which shows a marked advance in those mammals of higher degree that took to an arboreal life. Mention has been made already of this advance from Tree-shrew to Spectral Tarsier, from Lemur to Marmoset, from Monkey to Ape; and the advance can be reasonably connected with the assumption of the arboreal habit. For the fore-limbs became emancipated hands; there was a beginning of handling and holding, of manipulation and scrutinising; the snout shrank as vision dominated over the sense of smell; there was, as we have said, an increase in the parts of the brain concerned with visualising, with touch, with fine muscular movements, with attention, and so forth. There can be no doubt that the arboreal apprenticeship tended to help towards the improvement of brains.

The apes remained more or less arboreal, but the tentative men came to the ground. This step downward was really a stride upward. It meant being put to the test in a new set of surroundings; it meant a struggle with new enemies and competitors. There was more need for brains than for brawn; there was every need for the ancestors of men to stand by one another. It is quite likely that the descent from the trees was connected with geological and climatic changes—with the uplift of the Himalayas, near which the cradle of the human race may have been, and with the spread of drought or aridity. The apes found shelter in tropical forests and remained apes; man with his feet on the ground lifted his hands to heaven.

It has been suggested that the attainment of a more or less erect position was connected with the better evolution of the vocal organs, and it must have meant much when tentative men began to have not only words, as many animals have, but a language. A word is a sound which expresses emotion like anger, or indicates some thing, like food or enemy. But language means the expression of a judgment in a sentence. Socially imitated sounds, intelligible to others, made conversation possible, and they served as symbols of incalculable value in the art of thinking.

Finally there is a great truth in Rousseau's saying: Man did not make society; society made man. In the little groups of families which enabled tentative men to hold their own, there was an opportunity for more care both of the young and of the old, for more gentleness and good feeling, for more division of labour, for more freedom from worry. So man became Man. But we must inquire further into man's early inventions.

CHAPTER XXI

MAN'S EARLY INVENTIONS

THE story of Robinson Crusoe is very well told, and part of its evergreen interest is that it takes us back to man's making of things for himself, instead of getting them as ready-made gifts or purchases, as usually happens nowadays. But while Robinson Crusoe had in actual fact to begin afresh almost at the beginning, his head was full of lessons which he had learned at home. He was not like our earliest ancestors who had no memories of made things, but had to invent everything for themselves. It may not have been a comfortable time, but it cannot have been tedious—when genuine men began to think out inventions.

Let us try for a little to picture these early days when genuine men had appeared on the earth, but were still struggling hard to gain a firm footing. They had to find food, partly in the form of fruits and roots, and partly of an animal nature, such as eggs and shell-fish, besides the flesh of small beasts. They had also to make shelters against the cold of night and of winter, and against the storms of rain and snow, for it was only in certain places that caves were available. Then, again, many wild beasts, though not much given to attacking man—it is an exception for a lion or a tiger to be a " man-eater "—greatly resent intruders and will fiercely defend their haunts. So early man had to defend himself against wild animals. In many cases, perhaps, he ran away or climbed a tree. It was not till ages had passed that he began to hunt large animals actively. No doubt he learned many a useful lesson in caution, alertness, and presence of mind in avoiding a conflict, before he was in a position to show courage and wiliness in attack. Discretion came before valour.

When an explorer sets off nowadays he takes with him an equipment, which for the most part he buys from the stores, but our remote ancestors had either to find or to make everything for themselves. Our question is: *What were these early inventions?* One of the first would be a knife, a stone knife, with which to cut twigs and reeds for the shelter, or to take the skin off an animal, besides scraping the pelt and dividing the flesh. After a time the primitive knife would be used to sharpen a spear or to point a stake or to fashion a paddle. The first knives—stone knives—were probably *found*, but after a while they were *made*, though it was of course a very long time before they could be made of metal. The first *found* knives were splintered stones with sharp edges, especially flints which are sometimes formed when landslips and earthquakes break the rocks in pieces. In some places the teeth of sharks and whales thrown up on the beach made good first hammers, just as the finely barbed spines of the sting-ray would make small primitive saws. But when a man found a good natural knife or axe of flint he had not only an effective tool, but the means of making others. He could sharpen a straight sapling into a spear and he could chip other stones, even other flints, until they were fashioned into arrow-heads, spear-points, scrapers, adzes, wedges, and so forth.

Perhaps it was in chipping at flints that some men got their first idea of making fire—the fire that some of them must have previously recognised in the tree struck by lightning, or the flames of a volcano, or possibly when two dry branches of a tree rubbed hard and long against one another in the heat of the tropical sun. For it is said that a forest fire may sometimes arise in this way. But when a dust particle of flint is struck off by a very hard knock it glows with heat, and it is possible that primitive man caught this fateful spark in dry "touchwood," or in resinous dust from a tree which insects had bored. This was perhaps one of the ways in which he learned to make fire. It is not long ago since soldiers used "flint-lock" muskets, where the spark to ignite the gunpowder was obtained by striking a piece of flint very hard; and many people living to-day have seen a countryman lighting his pipe with a piece of paper or tinder set asmouldering by a spark struck from a flint. That he usually made the spark by knocking a sharp iron edge against the flint removed him in this respect only a short way from primitive man who had not advanced beyond knocking flint against flint.

There are other old-fashioned ways of making a fire. A hard stick may be rubbed backwards and forwards in a horizontal groove supplied with resinous dust, which is always available where there are fir-trees; or a vertical rod of wood may be turned round and round in a hole in a log till the sides become hot. This may be improved upon by using a coiled string to increase the rapidity of the rotation; and another contrivance is the "bow-drill." Some say that the Indian symbol called the "svastika" refers first of all to one of the devices for rapidly rotating a fire-drill.

When we strike a match nowadays we do not often remember that making fire was once a very laborious process, especially when the method was that of rubbing wood against wood. The amount of work that was required explains why care was often taken to keep the fire alight by means of smouldering touchwood or in other ways. It was one of the early housewife's duties to keep the fire from going out, and this was symbolised by the duty of the virgins in the temple of Vesta, who guarded the fire that was never allowed to die down. In some cases the fire was kept aglow in a vessel that could be carried from place to place, from one camp to another. How far away this seems from merely turning a switch,—but that is no more than two generations old.

How great an invention was fire-making! For without fire there could not have been any use of metals, not to speak of fuel; without fire there could not have been any cooking except in hot springs and volcanic ash; without fire many enjoyable parts of the earth would have remained practically uninhabitable by man; without fire man's knowledge and power would have remained sadly frozen!

We have spoken of the early invention of some sort of knife to cut with, but soon would come the use of string to bind with. This would be naturally suggested by the climbing and twining and tendril-bearing plants which are so common in forests, where those that are over-shaded often exert themselves greatly to reach the light. Grass-ropes are easily twisted, as hay-makers know; and the beautiful pendent nests of the weaver-birds would readily suggest baskets. Strips of skin and the tendons or sinews of large mammals would make stronger thongs, which would be used, for instance, to bind a stone axe firmly to its wooden shaft; and, as man became cleverer, to make a string for his bow.

Skins would be used very early for clothes and sandals, especially after man fashioned weapons and traps that enabled him to capture large mammals with tough hides. One must remember, however, that many simple peoples utilise pliable bark as part of their clothing, and that before there was true weaving or any spinning of thread there was plaiting of long leaves and fibres. Among simple peoples, who are in a way our " contemporary ancestors," there are still to be seen all possible stages in the history of tools and clothes—of everything, indeed, that is of vital importance to man. It is part of the task of the science of Anthropology to tell the story of the early inventions made by early man and, oftener perhaps, by early woman.

Another great step, the beginnings of which are hidden in the mist of the past, was cultivation. We cannot believe that man, when he became truly man, was very long of noticing Nature's sowing and what followed. The fruits fell from the trees and burst, liberating the seeds that sometimes sprouted. At the ripening time of year some pods and box-fruits explode, and jerk out their seeds under the drying up influence of the sun. In autumn long ago, just as to-day, the air was often full of parachuted fruits, like thistle-down, which the wind carries to and fro until they at last find anchorage. When man had a small clearing in the forest, he would bring home pleasant fruits, and the cast-away stones or seeds would sometimes sprout round the settlement. This might by and by prompt weeding to give the valuable plants a better chance. While the men were away hunting, the women started gardening. But we like to picture the exploring hunter getting very hungry and noticing the big kernels of the wild wheat, which is still to be found growing on the slopes of Mount Hermon. He would rub the grains in his hands, blow away the chaff, and enjoy a good mouthful, chewing for a long time. As he did this over and over again, he would make up his mind to *sow* this precious wheat nearer home. It was a prehistoric experiment of the utmost importance, for this or something like it was the first hint of a new kind of life—the agricultural. The cultivation of plants began in a small way.

But we must stay with the hunter a little longer, for it was no doubt he who made another great experiment, that of domesticating a young wolf. He had killed the parents, perhaps, and was a little repentant when he saw the playfulness and winsomeness of the cubs. So he took one home to please his own children, and what began as a pet became a partner—the domestic dog—the trusty guardian of his descendants' herds and hearth. The dog was the first animal to be domesticated, and it is interesting to notice in passing that one of the reasons why it attained to great heights of intelligence is to be found in the fact that it not only accepted man's tutelage, but became his responsible partner. In the same way domesticated horses have become much cleverer than domesticated cattle.

Another great invention was the boat. As Professor Tylor, one of the founders of Anthropology, says : " He who first, laying hold of a floating bough, found it would bear him up in the water, had made a beginning of navigation." It is one of the most beautiful of stories—the evolution of the boat. Let us picture the native astride on the log, so apt to turn turtle ; the shaping and balancing of the log ; the joining of log to log to make a raft ; the hollowing out of the log with knife and fire ; the use of a dug-out tree big enough to hold several adventurers ; the use of hides to make a skin-canoe ; the holding up of a mat or blanket to catch the wind ; the supporting of the primitive sail by a mast ; and so gradually onwards till there emerges before us the fishing boat—so important in opening up the world. Most of the big ideas of sailing craft were put into practice before recorded human history began !

An old device of much interest was wrapping the skinned mammal or the plucked bird in a casing of clay before it was put into the hot ashes to roast. This would improve the savouriness of the flesh and prevent burning. But the hardening of the clay probably suggested the beginning of pottery. The dome-shaped clay nests of the white ants are still sometimes used as ovens, and a big nest turned upside down and cleaned out makes a temporary tub. There are many kinds of natural cups and saucers—such as gourds, coco-nuts, pitcher-plants, clams, and other shells ; and some of these would probably serve as models to copy in clay.

Given a knife and a bison's horn, it was easy enough for the primitive inventor to make a spoon when he felt the need of one. Given a thong and a sharp borer of hard wood, he had in his hand the needle and thread of his descendants.

It must be admitted that some inventions are very difficult to account for, and well worth puzzling over. One of the most difficult is the wheel. It is easy to picture the early fishermen using felled trees as rollers on which to draw up the boat, especially when there had been good fishing. But the idea of the wheel is different from that of a roller. Perhaps a sawn section of a trunk would make a little wheel, whence big ones would follow. It is also possible that round stones from pot-holes in the river may have given an additional impulse towards a wheel, or the sight of the coiled up Rose of Jericho being driven by the wind across the desert! So far as we are aware, there is no hint of a locomotor wheel in the animal kingdom. Armadillos and hedgehogs bend themselves into living balls, but they do not roll along the ground; and no snake ever makes a hoop of its body!

We have spoken only of a few of the early inventions made long ago by our ancestors, but there are many interesting books, like Otis Mason's "Origins of Invention," and Tylor's "Anthropology," in which the study can be pursued, with, we are sure, increasing delight, for it is one of the most fascinating of stories.

CHAPTER XXII

THE MYSTERY OF THE MIND

WHEN we swallow a crumb we do not superintend the details of the process. It takes place, as we say, *automatically*, and so with coughing and sneezing, and closing our eye when something threatens to strike it, or drawing away our finger from a hot cinder. These, as we noticed in an earlier study, are called *reflex actions*, and they depend on linkages between certain nerve-cells and certain muscle-cells—linkages which are part of our inherited bodily structure. They are born in us. If we cross one leg over the other at the knee, and then get some one to strike the dangling leg firmly with the edge of his hand a little below the knee, our foot is jerked forcibly forwards and upwards. This " knee-jerk," which is used as a test by doctors, is a good example of a reflex action. But we must not overlook the fact, that, automatic as the action may seem, we are aware of it. This means that a message has been sent to our brain, our conscious attention is aroused, and we know what our foot has done. We did not will the movement of our foot, or control it, or attend to it, but we may be conscious of it ; and in this awareness there is a little glimmer of our " mind."

When we take some dry food like rice into our mouth, its presence provokes a flow of salivary juice, which contains a digestive ferment. The juice also lubricates the mouth, so that we are able to swallow the dry grains. This happens automatically, as we say. But it is interesting to notice that it used to be a custom in the Far East to give a suspected murderer some rice to swallow, and to judge him guilty if he could not get it down. The idea of this rough and ready justice was that if the man was guilty, he would be overwhelmed with fear, and the strong emotion or feeling,—*an activity of the mind to begin with*—would entirely hinder the secretion of the salivary juice. Thus the mind may be not only aware of the reflex action, *it may hinder it.* This is technically called *inhibition*, a word and a reality that we cannot dispense with. It is well known that many well-bred people are quite able to suppress a sneeze when it would be dreadfully out of place,—for instance on being presented at Court ; and a few people have their body so well under the rein of their mind that they show control much more wonderful than the suppression of a sneeze.

When a dog is shown a piece of meat " its mouth waters," and although the actual " watering." i.e. the flow of salivary juice, is involuntary or automatic or reflex, we must not try to shut the dog's mind out. For the dog *sees* the meat, it *knows* what the meat is good for, it has *memories* of the pleasure of eating meat, it has *feelings* in regard to meat, just as we have in regard to palatable things that we particularly like. If a mischievous person puts his teeth into a lemon when standing in front of a man playing a flute, the music has to stop, for the sight of the trickster biting the lemon produces such a copious flow of salivary juice in the player's mouth, that he cannot play. But an interesting fact should be noticed here, that the trick does not work when the musician does not know anything about lemons ! The success of the trick depends on the musician's previously enregistered experience of biting into a lemon *or some similar fruit*, so this brings us back to " mind," for the musician has a remembrance of the lemon, or a recognition of the significance of the lemon through his previous acquaintance with some other fruit that made his mouth water.

Now let us return to the dog that is shown a piece of meat. If a whistle is sounded whenever the meat is shown to the dog and afterwards given to the dog ; and if this lesson is repeated many many times, there is a quaint result. By and by the dog's mouth waters when the whistle is sounded, although there is no meat in view. For several weeks the saliva will not fail to flow when the whistle is sounded, although no meat is forthcoming at the time. A linkage has been formed, in the dog's nervous system, between the possibility of a pleasant mouthful and the sound of the whistle. The first stimulus, that is to say what pulled the trigger to begin with, was the sight of the palatable flesh ; but gradually a second stimulus—the whistle-sound—has become linked up with the effect of the first, namely mouth-watering, and so closely linked up that the sound of the whistle is followed by as much salivary juice as was induced by the sight of the meat. The second stimulus, which need not in itself mean anything, is not necessarily a sound ; the same result follows if, instead of sounding a whistle, a card marked in some conspicuous way is held up whenever the dog is shown the flesh. When we think of this from the bodily (physiological) side, we call it technically a " conditioned reflex," when we think of it from the mental (psychological) side, we call it an " established association." This is a difficult sentence, but it just means that all these doings of animals have to be thought of in connection with the mind, and also in connection with the body.

In the experiments referred to there was no connection, as we have said, between the sound of the whistle, and the meat, except that during numerous " lessons " the whistle was always sounded when the meat was shown. Similarly in our laboratory little groups of white mice learned in about forty twice-daily lessons to come running to the feeding dish whenever an electric bell was sounded, and after the association was firmly established they would continue coming over and over again although the cupboard was always bare when the bell rang. There is plainly no intelligible connection between the sound of the bell and the presence of the food ; yet the association becomes firmly established. We may say the same when the young chick learns to associate a particular kind of cluck on the hen's part with a tit-bit scraped out of the ground, and to connect another kind of cluck with danger overhead or close at hand. Sometimes, as we have seen in our study of Animal Behaviour, the association between a certain sound and a certain action, like the parent redshank's danger signal and the young bird's crouching, does not require to be learned at all, being inborn or instinctive ; but in many other cases the young creature builds up associations for itself. It must be noted, however, that in some of these cases in Wild Nature there *is* an intelligible connection between the signal and the action, and this means that " mind " may play its part in the learning. Thus the threatening appearance of a beast of prey, like a fox, may arouse fear in the heart of a young rabbit and pull the trigger of the instinct to take to its heels; but the situation is more intricate when the young otter stands stock still, or slips noiselessly into the pool, when it hears an unusual crackling of broken twigs. The clever and well-taught creature has learned to link together a suspicious sound and the appearance of a dangerous enemy. So it is with the human scout or explorer, except that his sense of the significant is much more highly developed. The more of this kind of behaviour there is, the more reason we have to speak of "mind."

Another familiar fact is that many kinds of activities which require very close attention to start with may gradually become *habitual.* It was difficult at first to ride a bicycle, we had to control so many movements at once, and very often the adjusting movements of our body did not come quick enough to prevent us from falling. But how easy it soon became; we learned to balance and steer almost, if not quite, automatically. So is it with playing some game of skill or some musical instrument; so is it with writing and with walking home through the crowded streets. A particular stimulus calls for a certain answer-back from our nerves and muscles, and when this happens over and over again the appropriate answer seems to come easier than anything else, and it is given without hesitation or attention or conscious control. A habitual action is often compared to a much trodden path along which we can move quickly. As the word "habit" is often used loosely for a way of living, as when we say that hedgehogs have very interesting habits, it is clearer to use the adjective *habitual,* and the long word *habituation* for the process of forming a habit in the true sense. The important point is that activities which required much intelligent control to begin with may become so engrained in the nervous system of man and animal that they cease to need more than a minimum of attention. The great advantage of this is that it saves time and thought; it sets the mind free to attend to something else.

Habits which are learned by the individual are not to be confused with other routine actions, discussed in connection with animal behaviour, which do not require to be learned at all, but are inborn, part of the hereditary equipment. These are called instinctive actions, and though they do not arise as habits arise, they have the same advantage that they do not require attention and control at every turn. A musician may play a difficult piece while he is thinking furiously about something else ; what he has done a thousand times has become habitual, though his mind probably has its hands on the reins all the time. But if a boy could play a difficult piece the very first time he tried, as a spider can spin an intricate web, that would be *instinctive*. Man has a considerable number of general inborn promptings, like self-preservation and kin-sympathy, but he has very little that really corresponds to the detailed instincts of, let us say, ants, bees, and spiders.

It may seem at first a strange thing to begin a study of mind by laying stress on the fact that in many activities the work of the mind is not very prominent. But the reason for this is that the science of to-day has shown us that much of everyday behaviour is the outcome of what has been enregistered in the previous experience of the individual or of the race. Many actions, that seem at first sight to require mind, have gradually become in great measure, if not quite, independent of mind. When we slip on a slide on the road we often recover ourselves very effectively and very neatly, and we may use a score of different muscles in so doing. But this recovery of balance is not in any sense intelligent ; it is an adjusting reaction connected with the semicircular canals of our ears ; it is born in us and almost ready-made, though it is partly developed unconsciously in the course of our youthful experience. Similarly when a cat is held upside down at a considerable height above a bed of straw and then let go, it *always* falls on its feet. Every " righting " movement that it makes is well known, most of the physiology of the useful adjustment is clear ; and, *as things are now*, there is no warrant for giving any credit to the cat's mind, alert as that mind may be.

Let us take another example at a much lower level in the Animal World. When young spiders are going to set off on an aerial journey, borne on the wings of the wind by parachutes of gossamer, as we pictured in our autumn study, the first thing they do is to climb up on posts or palings or tall herbs. From this vantage-ground, standing with their head to the breeze, they pay out long threads of silk on which they are soon borne away. How natural it is to credit these "gossamer spiders" with some understanding of the advantage there is in climbing up off the ground, for in this way they make a good start with a lessened risk of entanglement. In all probability, however, this is far too generous to spiders; for it is known that all young spiders, belonging to species that are not shy of the light, have an inborn obligation (or "tropism") to climb up, that is to move in a direction opposite to the pull of gravity. They do this until they cannot climb any further, or until some other influence bids them stop. The great French entomologist Fabre (d. 1916) tells us in one of his delightful books that a family of garden spiders, hatched out at the base of a fifteen-foot bamboo rod, began at once to climb up, and continued doing so, by slow stages, for four days, until they reached the top! The climbing-up has not, in the first instance, anything to do with gossamer-flights, and, *as things are now*, we may leave intelligence out of the question. Yet it is very useful in starting on an aerial journey.

We see, then, that many effective activities may be the outcome of what has become engrained or enregistered in the nervous system in the form of reflexes and tropisms and instincts, which do not require intelligent control from step to step. But care must be taken not to assume that the creatures showed no intelligence during the long succession of generations when the inborn promptings were being tried and tested, and before they took a firm hereditary grip of the constitution. We do not mean that the Garden Spider's instinctive weaving of a beautiful web was once an intelligently controlled series of actions, though some naturalists hold this view; but we mean that the suggestions to action, which arose from within, like reflexes, were put to the proof by the individual animal. It is likely that many inborn suggestions cropped up which did not justify themselves, and had to disappear, because, when put to the test of experience, they proved to be very risky or even fatal. This is a very difficult question, but the important general point is that many activities that seem at first glance, and at first thought, to demand intelligent control cannot be looked at in this generous way. They have become more or less independent of the mind. Many animals, such as geese and donkeys, are much cleverer than most people think; but there are many other animals, such as ants and bees, which are not nearly as clever as they seem.

Having made this admission, we may now ask why we attach so much importance to "mind," when so many effective activities can go on very well without its direct assistance. We mean by the "mind" the inner life of feeling and wishing and thinking, which cannot be described in terms of matter and energy. The brain is something we can weigh, but the mind is imponderable. The nervous system can be seen, the mind is invisible. The activity of the brain can be described in part in terms of chemical and physical changes, but these do not help us to describe our emotion as such, or our purpose as such, or our judgment as such. To use the two very useful technical words, nervous activity is *objective*, while mental activity is *subjective*. We have repeated the words "as such" because of the very important fact that mental activities are often very closely linked to bodily activities. Thus the emotion of anger, a mind-storm to begin with, excites various parts of the body, such as heart and lungs. In a very short time, by means of certain nerves the emotional excitement in the mind and brain influences a pair of ductless glands or "supra-renal bodies" which lie just above the kidneys. The hormone or chemical messenger, called adrenalin, which these supra-renal glands produce, is increased in amount by the nervous thrill and is distributed by the blood throughout the body, affecting the heart-beat, the breathing movements, the muscles, even the percentage of sugar in the blood, and so forth. It is in this way that a cat's hair stands on end when she is angry at an intruding dog. Our emotion or feeling includes not only a mental excitement, but the echoes of this excitement throughout the body.

Somewhat in the same way, our purpose, which is a powerful idea or an exciting image in our mind, may be accompanied by a flashing of the eye, a clenching of the teeth, a tightening of the muscles, a quickening of the circulation. Even our judgment or thinking is often accompanied by wrinkling our brows or half-closing our eyes or pursing our lips. Careful experiments have shown that there is a change of temperature when a student gets keen over a mathemetical problem; and there is a change in the electrical state of the skin when a student at an examination passes from translating Latin into English to tackle English into Latin. This is one of the biggest facts about the mind, that its activity is very closely linked to the body. Body thrills to mind, as when the angry man is prepared for a fight by the hormone of his supra-renal glands ; and mind thrills to body, as when vigorous health promotes quick and clear thinking.

Let us illustrate some of the ways in which " mind " and " body " seem to be bound together. We mean by " body " the visible, weighable, living creature, its nervous system in particular. We mean by " mind " the inner life, invisible and unweighable, including in our own case all our feelings thoughts, and purposes ; and we should never lose hold of the fact that it is through our minds that we know anything at all. It is in our mind that the outside world is mirrored and measured; it is through our mind that we have come to understand a little about our body and brain. Therefore we can never succeed in " explaining " our mental activity as an outcome of the matter and motion in our brain, for it is through our mental activity that we have built up all that we know about brain and nerves, matter and motion, electrons and protons, ether and radiations. When you think of it, there is a contradiction in trying to "explain" mind in terms of anything material that our mind has disclosed.

Yet our mental life is real and our bodily life is real, and what we are trying to do in this study is to begin to think of the relation between the two, for this is the unsolved riddle or mystery of mind. We know that it has been a riddle for all the centuries since men began to record their thoughts ; and it remains a riddle still.

One of the biggest facts in the history or evolution of our world is that as the ages passed, millions after millions of years, there was more and more evidence of " mind " among animals. As the brains of animals became larger and, what is as important, more intricate, there was a growing freedom of mind. That is to say, animals became more intelligent, more full of feeling, and more determined in their purposes. So there seems to be a direct connection between the complexity of the brain and the vigour of mental life. Some considerable complexity in the structure of the nervous system is necessary if the mind is to get free play.

In a new-born infant there is not much sign of mind, yet

the infant may become a genius. But as we cannot think of the mind coming in from without in any mysterious way at any stage whatsoever, it must be there from the very beginning of the individual life. It must be somehow present in the fertilised egg-cell, and it makes no difference whether we are dealing with a bee or a bird, a dog or a child. No doubt the mind owes much to the way in which it is nourished and educated, but somehow, though we cannot understand how, the germ of the mind is present from the first, just as the germ of the body is. Both are like buds which have to be unfolded or developed ; and what corresponds in the care of buds to the sunshine and the rain, the wind and the morning dew, is represented in our own early development by all that is included in the word " nurture " —food, fresh air, sunlight, warmth, exercise, home, parents, school, friends, play, work, books, pictures, and the outside world. This " nurture " is necessary for the development of the body and also for the opening of the blossoms of the mind ; but our present point is that as the brain of the young child becomes more intricate in its structure, there is more opportunity for the mind to express itself vigorously. The development of the mind goes hand in hand with the development of the brain, indeed with the development of the body as a whole.

It sometimes happens that the development of the body goes wrong, and this handicaps the growth of the mind. This is well illustrated when something hinders the proper activity of the ductless gland that is called the thyroid. It lies on each side of our voice-box or larynx (Adam's apple), and although it is very small it is essential to the continued health of body and mind. It manufactures a hormone called thyroxin, which the chemist can now build up artificially ; and this thyroxin, distributed by the blood through the body, regulates development and that harmonious working-together of parts which spells health. If the child's thyroid is not working rightly, there is an arrest of development, and the child remains infantile in body and in mind. But one of the triumphs of modern medicine is the discovery that the child can be rescued by giving it thyroid treatment, that is to say by introducing thyroid extract or artificial thyroxin along with its food. The potent chemical substance works like magic and may compensate for the deficiency in the supply which the child should normally make for itself. But our present point is that a defect in the internal working of the body may handicap the mind in a terrible way. The development of the healthy mind goes hand in hand with the development of the healthy body. *Mens sana in corpore sano.*

What is true of the child is true also of the old man ; the fatigue of part of the brain seems to dull part of the mind, and an injury to a particular region of the brain may be followed by a particular handicapping of the mind. However we think about the two sides of our life, the bodily and mental, the nervous and the psychical, the fact is certain that they are closely linked together.

A clot on the brain, caused perhaps by the bursting of a small blood-vessel, may make the whole mind dark ; so we should try to take good care of our brains as well as of our minds. Something wrong with the eyes or with the digestion may blur a great man's outlook on life or change his good temper into peevishness. But we should never forget the other side, that a robust mind often triumphs over a sickly body, that resolution sometimes drives away disease, and that the habit of happiness often banishes fatigue. " A merry heart is the life of the flesh." *No one yet knows how much our mind may do for our body.*

But the difficult question remains : How are we to think of body and mind in relation to one another ? There are two main views, each with its merits, each with its difficulties. Many wise men think that the mind is a distinct reality which plays upon the body, as a musician on his violin. The mind needs the body to keep it in touch with the outside world, to enable it to get things done, and to help it to express itself ; but it is as different from the body as the musician is different from his instrument, and it is the greater of the two.

But there are other wise men who think that the living creature is a unity with two aspects, like the two sides of a shield, or the inner and the outer surfaces of a dome,—two aspects which are inseparable. At one time the activity of the living creature seems for the most part bodily, as when it is digesting its food. But every one knows how good news received by the mind helps the digestion that is going on in the food-canal. So even in the process of digesting, the living creature, such as ourself, is a *mind*-BODY, if we put the prominent side in capitals. At another time, however, we are thinking hard, and this seems of course the business of our mind. Yet our thinking is influenced by the state of our body, by its health, or its half-health, or its disease; and so the philosopher, wrapped up in his thoughts, is a *body*-MIND. The puzzle is too difficult for us, but the big certainty is that the mental life and the bodily life are both real, though so closely interlinked. Both are sources of delight and of danger; both are to be looked after as carefully as we can. For millions of years the part played by the inner or mental life in animals has made headway, becoming more and more important; there has been a growing freedom of mind. We should be very careful not to suppose that this emancipation of mind has come to a standstill.

CHAPTER XXIII

THE BEGINNINGS OF SOCIETY

THERE is an old saying, " The proper study of mankind is man," but it would be just as true to say, " The proper study of man is mankind," if we mean by mankind all the human societies. For the highest outcome of evolution as yet reached on our earth is a human society, imperfect as all of them are. So, as we began our studies with a distant nebula, we end them for to-day—to begin at a higher level to-morrow—with human society. But we cannot do more than say a little about its beginnings long ago.

Let us start with *animal* societies, such as are illustrated by ant-hills and bee-hives, beaver-villages and packs of wolves. An assemblage of many animals of the same kind does not make a society, for a rabbit warren does not deserve that name, nor the multitude of mites in the huge cavern of an old cheese. The distinctive feature of a society is that the individuals can act together as a unity, combining their efforts in defence, or in attack, or in work, or in some common enterprise. A society implies some degree of corporate life, when the members act coherently and harmoniously as a unit,—when the whole is more than the sum of its parts! Thus the Amazon ants may combine in a slave-raid, or the beavers may combine to dig a canal through a large island in the middle of a river. There are many different grades of animal sociality, thus the rooks are much more socialised than the parrots, gregarious as these often are. There is much more concerted action in a pack of wolves than in a herd of wild horses. But whether the social note is loud or faint, if it is there at all it implies some self-subordination of the individual to the interests of the community. The contrast is with the solitary, self-contained, independent, each-for-himself animals, which are admirably suited for certain kinds of life, and may be also most admirable in their parental care and in their monogamy, in their courage and in their resourcefulness. The contrast between a communal or social régime and an individualist or solitary mode of life is not a contrast in morals; it is a contrast in ways of getting a living and in keeping a firm foothold in the struggle for existence.

Another general feature of the social mode of life, when it gets beyond mere gregariousness, is some alleviation of the *individual* struggle for existence. In diverse degrees the society serves as a shield to the individual. This means that certain types of individual that could not survive alone, may survive under the society's shelter. Among the honey-ants of Texas there are individuals which are utilised as honey-pots, but that would be impossible except in a society! Among the White Ants or termites the big-jawed soldiers cannot gnaw wood as the workers do, so they are fed in return for their more or less effective military services. The drone-bees in a hive, though very energetic in flying about, have lost the habit of foraging, and get their food as members of the society or big family. The difference between a society and a big family can hardly be upheld.

Some animal societies are *mainly* on an intelligent basis, as among monkeys, horses, cattle, elephants, beavers, and rooks, while others are *mainly* on an instinctive basis, as among ants, bees, wasps, and termites. But the contrast must not be pressed too hard, for beavers are sometimes children of instinct and bees show occasional flashes of intelligence. Human societies are partly instinctive, mainly intelligent, and occasionally rational.

Perhaps we shall better understand the beginnings of human societies if we first inquire into the advantages of social life among animals. (1) Many small animals, such as ants, insignificant in themselves, afford good illustrations of the adage that union is strength. (2) What an individual ant or beaver could not accomplish may be achieved by concerted action, as when ants combine to bring a large victim to the nest, or when beavers unite in cutting a canal. (3) Energy may be economised in a community, especially when there is division of labour, as is so frequent among ants and termites. This division of labour sometimes goes so far that parentage is mainly restricted to certain individuals, notably the queens and drones of the bee-hive and the ant-hill. (4) We cannot help thinking that it must have meant a good deal in the evolution of an animal society when there was something in the way of *permanent products,* such as communal shelter, or store, or camp. An ant-hill, a bee-hive, a termitary, a beaver's pond, must be regarded as the beginning of the social heritage which has meant so much in man's case. (5) Finally, there must surely be some degree of kin-sympathy in every animal society—a " social atmosphere " in which the mental and moral qualities have perhaps more chance to express themselves than in the solitary mode of life. This would be most marked in the case of societies on an intelligent basis, for instinctive societies are apt to become too stereotyped. Communities on an intelligent basis are likely to foster the growth of wits and kindly feeling, as well as the anticipations of language and art.

Why are there not more instances of social animals ? It probably requires a considerable degree of kin-sympathy and brain-intricacy. Mites could not form a society. There must also be a power of rapid multiplication, for a small society is almost a contradiction in terms. Moreover there are some ways of living that preclude concerted action, as in most forms of hunting and fishing. There are only two or three instances of sociality among spiders. These and other reasons help us to understand why the list of social animals is not longer. The formation of a society makes certain demands on its members,—above all some degree of self-subordination ; and these demands cannot always be met. Successful as social life is, it is on the whole for the elect.

Perhaps we shall be helped to understand social life in mankind if we linger a little longer over social activities among animals. There may be concerted action or communal enterprise, as in making a beaver-dam. There may be combination in defence, as when wasps unite against an intruder, wild cattle against the menacing carnivore, rooks against a hawk. Or the union may be aggressive as when the wolves in a pack unite against a deer, or weasels against a dog. There may be co-operation in food-getting as when pelicans in a half-circle close in upon fishes in the estuary ; or combined action in making a shelter or a store, as in the termitary and the bee-hive.

But the social activities of animals are sometimes quaintly

subtle. Thus among some of the slave-keeping ants, the slaves may assist in capturing others like themselves. Many migratory birds fly in a wedge-shaped formation, which lessens the physical exertion and devolves the responsibility of guidance on a leader, who can be changed when he or she gets tired. Corporate nesting is illustrated by the Republican Birds that make a huge composite erection almost smothering a tree; and there are some social features in the gregarious nesting of rooks. Social activities also include the "wars" of some species of ants, the ploys and plays of others, the drilling manœuvres of the penguins and their games, the choruses of some joyous birds, and the "community singing" of the Howling Monkeys.

Another line of social activity is the evolution of means of communication, by sound-signals in particular. The first use of the voice was as a sex-call, and this use remains prominent in many vocal animals, such as the croaking frogs. A second phase in the evolution of the voice is represented by those animals in which the young ones call to their parents—the unhatched crocodile pipes from within the egg-shell—or the parent's call to the young ones, as when the partridges utter the danger-cry that makes their offspring squat and lie still. Later on in the evolution the sounds become kin-signals, which sometimes save a difficult situation. Thus an isolated monkey attacked by an eagle may summon its kindred and entirely alter the crisis. Gradually there came to be "words" by which we mean sounds associated with particular provocations, feelings, desires, or even objects. Rooks, dogs, monkeys, and many other brainy creatures have many "words," but they never make a sentence. There is no true language until socially imitated sounds are used to express a judgment; and even chimpanzees, which have a large vocabulary, never do that. Man has a monopoly of language, though many animals have speech and words. But our present point is that a society favours the development of means of communication. These need not be vocal, for bees get news from one another by smell, and in many mammals there are gestures as well as sounds. When two ants stroke antennæ there seems to be a mingling of touch-tidings and smell-tidings.

Of great interest in the social life of animals is the gradual appearance of customs and conventions, what might be called " folk-ways." These are based partly on the engrained promptings which have become part of the racial inheritance, partly on the apprenticeship that the animal may have to serve (as in worker-bees) to the traditional routine, and partly, no doubt, to the established division of labour and to the presence of permanent products such as the termitary of the white ants with its intricate internal architecture. Why does an ant seem forced to behave in such and such a way ? In trying to answer, we may perhaps distinguish an internal hereditary compulsion, which prompts, for instance, the feeding of the hungry, and an external or environmental compulsion imposed by the nature of the home, the particular form of the quest for food, and the established framework of the society, as expressed, for instance, in the division of labour. Thus, as we have mentioned already, the soldier white-ants *must* be fed by the workers, for they cannot feed themselves.

In societies of fine-brained animals there may be the beginning of something like social compulsion, something pointing onwards to public opinion in mankind. Perhaps, to take a familiar instance, there is some expression of this in the cold-shouldering of the drones in a bee-hive. It seems to grow in intensity and it ends in their massacre towards the end of summer.

But let us turn now to the beginnings of *human* societies. Here we must expect to find great differences and yet great samenesses. What are the great differences ? In what ways does even a humble human society rise high above a bee-hive or a beaver-village ? We often hear about " the human hive " and " the instinct of the herd," but these phrases tend to exaggerate the resemblances and slur over the differences. Man has language, making sentences, expressing judgments ; animals never rise above words. Man has reason, that is to say the power of working with general ideas ; animals are not known to rise above intelligence. Man has in varying measure an awareness of his own history, but that is beyond the animal except in so far as engrained promptings form part of their inheritance. Man has much more in the way of a social heritage—traditions, customs, institutions, laws, literature, and art. There are hints of something of this sort in an ant-hill that lasts for generations, but man in his ascent has come to be more dependent than any animal on the heritage outside himself—the social heritage—and he has almost unlimited powers of making this better and better. Finally, man has the power, if he would only exercise it more, of guiding his conduct in reference to ideals. Animals are often kind parents, affectionate lovers, helpful to their kindred, devoted to their society, but there seems no warrant for supposing that they ever think of their duty! They may be good and kind, but they do not know what "ought" means. Perhaps we may draw a distinction between animal *behaviour*

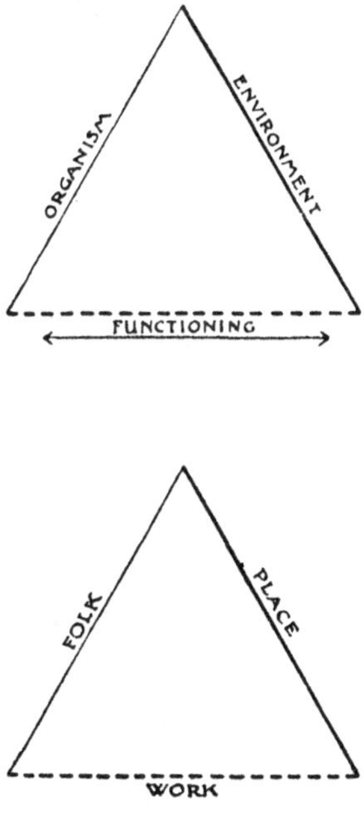

FIG. 28.—BIOLOGICAL AND SOCIOLOGICAL PARALLELISM.

The three biological aspects or categories—the organism, its functioning, and its environment—correspond to the three sociological aspects—Folk, Work, and Place. Using the first letters, we may say that O : F : E corresponds to Fk : W : P. The three corresponding lines are indicated in similar ways.

and human *conduct;* but it is not always that man rises above behaviour.

So much for differences, but let us not forget the resemblances between animal and human societies. It used to be customary to compare a human society with an animal body, and to speak of the "social organism"; thus the government was compared to the nervous system, and the workers to the muscles, and transport to the circulation of the blood, and so on. But that was a confusion of thought;

the true comparison is between a human society and an animal society. We have seen that the comparison must not be pushed too far, and another reason for being careful is that the members of an animal society are all nearly related, whereas in a human society this is usually far from being the case.

The science of societies is SOCIOLOGY and the sociologist

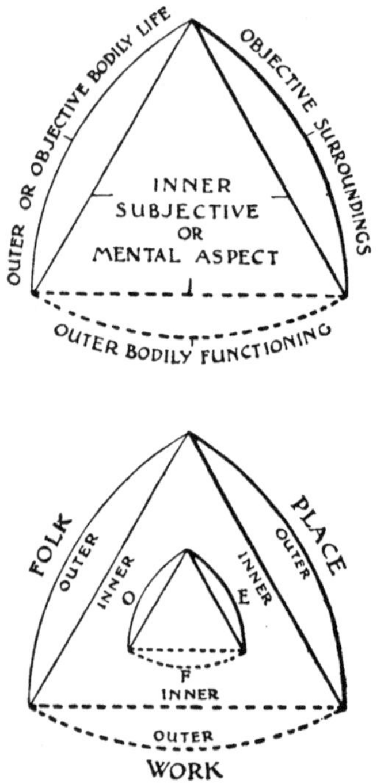

FIG. 29.

The upper figure suggests that there are subjective or mental aspects of the organism, its functioning and its environment. This upper figure is then placed inside the Folk, Work, Place diagram, where again the inner and the outer aspects have, of course, to be distinguished. The three kinds of lines will be noticed—thin, thick, and broken. O, F, E; the first letters of organism, functioning, and environment.

always seeks to describe the life of the society in terms of FOLK, WORK, and PLACE. Everything depends on the people themselves, what they busy themselves with, and what their surroundings are. Now these three ideas—Folk, Work, Place, which the French sociologist Leplay called *Famille, Travail, Lieu*—obviously correspond to what in biological language we call Organisms, Function, and Environment. These three fundamental ideas apply to all living creatures, and thus there must be deep resemblances between animal and human societies. For men are organisms functioning in a particular environment, and thus they resemble animals though they rise high above them.

If we believe that tentative men, who might be called Hominids but not Homo, arose from a stock common to them and to the Anthropoid Apes, we are linking mankind back to creatures that are nearer the social than the solitary. Gorillas sometimes prowl about in small troups, and chimpanzees are very fond of one another's company. Although it cannot be said that any of the Anthropoid Apes live in communities, they are related to monkeys, among which the social note is often sounded. We are not of course including any living monkeys in man's lineage—no one believes *that*—but they represent a kindred stock, specialised for arboreal life, and they are often more than gregarious. They sometimes combine to drive off intruders; they will tear an over-venturesome eagle to pieces; they sometimes unite to raid an orchard; and they can execute a retreat, not in a disorderly rout, but with some hint of tactics and care for the young.

It cannot be supposed for a moment that a human society grew out of a troup of gorillas. Man made his own society. But our point is that man is solidary with mammals, among which the social note is often sounded. Think of beavers in their "village," prairie dogs in their "town," wolves in their pack, elephants in their troup, wild horses in their herd, dolphins in their "school." It is true that many mammals follow the each-for-himself line of life, but sociality has also its numerous illustrations, and it is, so to speak, in the blood of the class to which Man zoologically belongs and from within which he certainly emerged. Our point is that if all mammals were like the cats that walk alone, the emergence of social man would be much more of a scientific puzzle than it is.

There is much truth in Rousseau's saying: Man did not make Society; Society made Man. For it was in society that man's characteristics—such as reflectiveness, language, and gentleness—would have most chance of surviving. That is to say, new departures or variations in the direction of these estimable qualities would be most likely to be fostered in social conditions. A very clever creature might evolve as a solitary, but can we picture man's emotional and ethical and artistic evolution apart from a social heritage? In this sense society made man.

But how did the human society begin? It is not improbable, as Professor Barell first suggested, that Man and the Himalayas arose simultaneously, towards the end of the Miocene Period, over a million years ago. Sir Arthur Smith Woodward tells us that "as the land rose, the temperature would be lowered, and some of the apes which had previously lived in the warm forest would be trapped to the north of the raised area." As the forest shrank and gave place to plains, the ancestors of man had to face living on the ground. If they had remained arboreal, or semi-arboreal like the apes, there might never have been men.

Our theory is that the early forerunners of Homo were forced to try a new environment, and that this was a good reason for their standing by one another socially. They were doubtless big-brained, and there is no reason to suppose that they were lacking in courage or resourcefulness, but they were adopting a new rôle as terrestrial creatures, with formidable wild beasts as competitors. Their hope was not only in their wits, but in their solidarity. Union is strength.

Another reason for clubbing together may be found in the prolonged human infancy, with its appealing helplessness, for this involved self-subordinating division of labour. For an isolated human family the struggle for existence was too keen beyond the sanctuary of the trees. Thus arose a self-preservative linking of families into simple societies.

But it may be asked, Does not an otter family and many another animal family stand alone? Why was society so necessary for man? Part of the answer may be found in the otter's remarkable fitness for an each-for-himself mode of life. It is endowed with great vigour, nimble wits, a mastery of mountain and moorland, river and sea, a roving disposition, masterly resourcefulness, and a capacity for thriving on very varied flesh-food. But the ancestors of men were trying a new haunt, they had more brains than brawn, they had at first no chance against lions and tigers except by outwitting them, they inclined to be gentle of heart, and, as we have said, their children were for an unusually long time helpless. Thus the pre-men, no doubt in obedience to their engrained promptings in favour of mutual aid and sociality, found safety in simple societies.

In face of great difficulties modern man often sits down and reflects on ways and means, arguing to himself that this or that change might save the situation. But we must not read ourselves back into our very distant ancestors, a million years ago, who made the first experiments in society-making. They did not combine their families because they foresaw possible advantages. They obeyed their social promptings and then discovered more or less dimly that there was strength in their weakness. And whenever the simple society began to justify itself in giving man a firmer foothold in the struggle for existence, variations in the direction of increased sociality would tend to survive. As we have noticed in regard to animals, so in man's case, the society would achieve more than the isolated family; precious individuals, such as pioneer thinkers and artists, would have a chance to survive under the society shield; life would be more secure for the pioneering children and for the aged who treasured the lessons of experience; conversation would become a habit and men would compare notes around the fire. The early societies were small beginnings, but even in the smallest there was the promise of a great future still beyond our reach, though not beyond our hopes.

Our outline-story of the steps leading to human societies has brought us to our early forerunners, who were no longer tentative men, but true men of our own *Homo sapiens* species. We picture them living in small communities, sometimes in clustered dry caves, sometimes in rough-and-ready shelters near the edge of a great forest, sometimes in tents on luxuriant grass-land. We must credit them with having learned how to light a fire and how to guard it. They had chipped unpolished stone implements, which they used with increasing skill in fashioning weapons for the chase, in preparing their food, and in making rough clothes. They lived on the fruits of the earth, and on the small animals of the woods, and sometimes of the seashore. But they had as yet no pottery, no metals, no agriculture, and no domestic animals—not even the dog. This was in the Old Stone (Palæolithic) time, towards the end of the fourth and last of the Quaternary Ice Ages, perhaps ten thousand years ago.

A surprising amount of reliable information has been obtained from tell-tale remains found buried in the gradually rising floors of caves and in the gravels beside old encampments. It seems safe to say that these early forerunners of ours were for the most part hunters, at first of small and afterwards of larger animals; that they were not very quarrelsome, but chiefly pre-occupied with getting food and with holding their own against animal competitors; that they were not quickly inventive, for the same type of tool sometimes lasted for a thousand years (very different from the modern pace!); and that some of them had notable artistic skill, as shown, for instance, in the spirited drawings of bison, deer, mammoth, and the like on the walls of the fire-lit caves.

There are some reasons for believing that the cradle of mankind was some plateau land in Central Asia, not very far from the great Plateau of Tibet. From this uncertain ancient home-land men spread over the world, and there is no mystery about the general fact of dispersal. Trekking is a common feature among higher animals, prompted sometimes by overcrowding and sometimes by a change in climate or in the surface features of the earth. In man's case, more than in any animal's, there would also be the spur of curiosity and adventure—a spur that became sharper as the ages passed.

Many of us know upland tracts among the hills in which three or four great rivers have their head-waters, and from which they flow in opposite directions and into different seas. The mouths of these rivers may be hundreds, even thousands, of miles apart. So we must think of streams of human life flowing in different directions from the Asiatic plateau. Thus the Semite stream may have flowed to Arabia, the Hamite stream to the Mediterranean region, the Negroid stream to Africa, and other streams eastward to Mongolia, south-east to Australia, and west, along the steppe-country, to Northern Europe. Partly as the expression of different temperaments and partly in response to the varied conditions of the country, there arose the primitive occupations, which have had far-reaching influences on human history, thought, and art, namely the hunters, the herdsmen, and the tillers of the soil, to put them in their order of emergence. It is evident that the hunters and herdsmen would be more restless or nomadic than those who began agriculture; so it is to the last that we must look for the origin of more or less lasting settlements. Among the early herdsmen began the domestication of animals, the dog first of all, for without this partner a herd is scarcely possible. Among the early tillers of the earth, who sought to secure the more abundant growth of useful wild plants, there was the beginning of cultivation in the stricter sense. We must not think of the different occupations as the same in different countries, or as following one another in linear order; they must have been very diverse according to what the country afforded, and there must have been great variety in the times at which they became well-marked in different peoples. Another primitive occupation, prompted by a rich seashore or a river abounding in salmon was of course that of fishermen, who were sometimes the predecessors of mariners.

As we picture the early human trekking and its persistence for many hundreds of centuries, we must bear in mind that a new force had appeared on the earth, not merely a creature who could pit wits against muscle, for this many of his precedessors had tried to do, but a creature who had reason as well as intelligence to help him, and who began very early the habit of registering his advances outside himself. We mean by reason the ability to work, play, and experiment with general ideas or concepts, and we mean by external registration everything of the nature of a tradition, a record, a permanent product,—gradually growing into literature and art. Even a stone-circle, arranged in reference to the seasons, was a great event in pre-history; and a calendar was still greater.

When wild sheep in their search for new countries found themselves in cold surroundings they would probably answer-back by growing thicker and longer fleece. This individual modification has been proved in the case of domesticated sheep. If the lambs of the wild sheep survived the cold they might grow up to have still thicker and longer fleece, for they would be subjected to the tonic influence of the cold for a longer time than their parents, who were adults, we are supposing, when they arrived in the cold country. This also has been proved in regard to domesticated sheep. But as no cases are known of further increase in the fleece in subsequent generations, it does not seem likely that evolution worked in this direct way. It is more likely that variations or new departures cropping up from within, in the usual half-mysterious way, formed the raw materials for sifting. That is to say, variants in the direction of thicker fleece would be favoured, e.g. in having longer lives and more offspring, while variants in the direction of thinner fleece would be winnowed out or eliminated. It is in some such way that most zoologists try to account for such animals as the Yak, a small ox with an extraordinarily hardy constitution, well fitted for the uplands of Tibet and adjacent regions.

The same processes of varying, entailing, and sifting must have operated when the prehistoric explorers came under the influence of diverse climates and surroundings; but we wish to emphasise the obvious point that man with his plastic reasonableness gradually became able to save the situation by artificial devices, such as clothing and shelters. His answers-back to climate were less restricted than the animal's. But in any case we must allow that climate was an important factor in human evolution. In northern countries, for instance, there would be a call for qualities which would not be readily evoked in easy-going tropical conditions.

Among our early forerunners who were mainly hunters, there was not much difference of occupation, except between man and woman; but as wandering gave place to settling there were growing opportunities for that division of labour that has been such an important factor in the evolution of human society. Thus there was a gradual appearance of tool-makers, clothiers, builders, gardeners, potters, artists, and so forth. There would be a great increase in this division of labour when man took one of the greatest steps in his history and began to work with metals—gold, copper, bronze, and iron. Authorities tell us that bronze was invented in northern Persia between 2500 and 2000 B.C., and in the tomb of Tut-Anth-Amen (1353 B.C.) they found an iron dagger.

We cannot continue our story of the evolution of societies, but we cannot shut our eyes to another factor which has been the cause of much change, *though perhaps not of much progress*. We mean the direct conflict of races in raids and wars, which still continue. Of the value of this form of costly struggle the common sense and common conscience of civilisation may be said to have begun to be very suspicious.

It seems certain that man cannot make progress without

some kind of sifting. The early sifting was effected by wild beasts, by the forces of Nature, by hunger and cold. Later on, there was sifting inside the society, when the rebellious, the lazy, the anti-social, and the weakly were eliminated—often very roughly. There was on the other side the positive favouring of the energetic, the vigorous, the sympathetic, and those who were quick to use the social heritage, which grew from millennium to millennium. By and by there appeared another sieve, often very indiscriminating, namely disease; it became a formidable sieve as crowding and artificiality of life increased.

In modern times man has made determined endeavours to throw off Nature's sifting—in such forms as famine and flood, plague and weakliness. Intolerant of every form of rapid elimination, he has rebelled against the older forms of winnowing. This has had undoubtedly good results, as in the conquest of many diseases and the more successful and economical use of Nature's resources. But the danger is lest man fail to replace Nature's sifting or selection by some subtler and more rational social sifting, which will prevent the tares from smothering the wheat. As George Meredith said:

> Behold the life of ease, it drifts.
> The sharpened life commands its course;
> She winnows, winnows roughly, sifts,
> To dip her chosen in her source.
> Contention is the vital force
> Whence pluck they brain, her prize of gifts.

CHAPTER XXIV

EVOLUTION

ALL through our studies there runs a big idea—the idea of evolution. This means that everything is the outcome of a long natural process of *Becoming*. Wherever we look we find that the present is the child of the past and the parent of the future. The rock-records tell us plainly that the present-day one-toed horse began to evolve from extinct ancestors, which had four toes in front and three behind, and were only about a foot high. It is certain that birds evolved from a stock of extinct reptiles, and reptiles from amphibian ancestry, and amphibians from fishes. So it is all along the line, the animals of to-day have behind them a half-known pedigree which leads back and back to simpler and, as we say, more " generalised " ancestors. The only exception is in the case of those living creatures that have degenerated or retrograded in connection with a parasitic or a sedentary way of living. For their ancestors would be higher not lower on the scale. Evolution has been on the whole progressive, but in certain cases it has been retrogressive. In other words, while evolution has usually been onwards and upwards in its sublimely slow changes, it has sometimes slipped backwards and downwards.

But we have to face the difficulty that the word evolution is rather *overworked*. It is used in reference to so many different orders of fact—the solar system, the earth's scenery, the plants and animals, the climates in different ages, the chemical elements, the societies of animals, man himself, his language, institutions, art, customs, and ideas. We even speak of the evolution of evolution theories! In many cases these different kinds of evolution have not much in common except that they show *a process of Becoming*.

Perhaps the best way out of this difficulty is to use the

word "evolution" with a qualifying adjective. Thus the process by which a great nebular mass gave rise to a system of stars, or by which an enormous sun whirled off planets, might be called *cosmic* evolution. It plainly differs from evolution among living creatures, for there is nothing corresponding to successive generations, and there is no process of sifting or selection, in the course of which some forms survive while others are wiped out.

If we turn, in the second place, to the earth itself, we know that it too has had a history, which is often called its evolution. It was once a revolving gaseous mass with a very high temperature; it cooled and it shrank; it formed a crust round a core; the crust was buckled and scrapped, powdered and consolidated again; continents were separated from oceans, and so on. The geologists speak of the evolution of mountains, minerals, scenery, and climate. When we take a long view we may, perhaps, be inclined to say that our earth made *progress*, for it gradually became fit to be a cradle and a home of life. It may be that the old-fashioned word "genesis" would be a useful one to apply to the Becoming of the earth. An alternative would be to use an adjective as prefix, and speak of *terrestrial* evolution. But we must be clear that the material composing the earth has remained much the same in amount from the beginning until now; there is no multiplication of earths, no struggle among competitors. The moons of Jupiter do not struggle amongst themselves.

In the domain of chemistry there are, as we have noticed, processes by which one element may give rise to another. The dream of the alchemists has come true, though not in the way they expected. Thus uranium may give rise to protactinium, which produces actinium, which produces lead. Or uranium may give rise to ionium, which may give rise to radium, which also gives origin to lead. About eight different transmutations seem to end in that rather dull stuff—lead! At present, on the earth, all these radio-active changes are proceeding in the same direction—towards simplification. The chemical clocks are all running down. But it is possible that they are being wound up in the stars, and it may be that the simpler elements were compounded into the more complex in ages long since past. As the (92 — 4) different kinds of elements only differ from one another in the number, disposition, and behaviour of their component electrons and protons, it is natural to think of an evolution of the different kinds of matter. This might be called *chemical evolution ;* or the old word *transmutation* might be conveniently applied to the changes by which certain chemical elements are known to give rise to others.

In appropriate conditions there is a widespread tendency for units to draw together, for "pickles" to become "mickles"; and we cannot but think of a very fundamental kind of evolution by which electrons and protons became those minute constellations that we call atoms, and groups of atoms formed molecules, and clusters of molecules became colloidal micellæ (particles and droplets) in a fluid medium. Except in the last case, the formation of colloids, there is little indication of the continuance of up-building in our non-living surroundings on the earth ; but we cannot forget the most important vital process in the world, the process of *photosynthesis*, by which the green leaves build up carbon-compounds, like sugar, from carbon dioxide and water ; and here we must also take account of the wonderful things the clever synthetic chemist is able to do by building up carbon-compounds in his laboratory. Out of relatively simple materials he manufactures such subtleties as indigo, salicylic acid, and adrenalin. As he often makes things which are complete novelties, not found as such in Nature, e.g. artificial perfumes, drugs, and explosives, he is not undeserving of being called " creative." Perhaps this kind of evolution is best called *chemical synthesis*, with the adjectives " natural " and " artificial " prefixed if need be.

Here it is interesting to recall the fact that one species of plant or animal may differ from its near relatives or cousins in having some different protein (a complex nitrogenous carbon-compound like white of egg) in its protoplasm, so that the syntheses effected by the creative chemist in the laboratory are not altogether remote from the changes that occur when a plant or animal gives rise to a new departure or variation—the first step towards a possible new species. Moreover, as we mentioned in a previous study, a series of different species, as among roses, may differ from one another in the regular increase in the number of their nuclear rods or chromosomes. Thus among species of roses, to take a fragrant instance, there is a series of different kinds with fourteen, twenty-eight, and fifty-six chromosomes. It is a misleading simplicity to try to whittle away the differences between the living and the not-living; but it is a mistake in the opposite direction to shut our eyes to the fact that some of the changes in living creatures are *in a line with* changes in not-living materials.

There is no excuse for mixing up with evolution what we have discussed in our Nineteenth study—the *development* of the individual. Tadpoles develop from the eggs that are liberated by the female frogs as "spawn," and the tadpoles go on to develop, somewhat circuitously, into little frogs. But this is very different from the racial history that began at least 100 million years ago in the time of the Old Red Sandstone, when the first Amphibians emerged from a stock of old-fashioned fishes. The evolution of Amphibians has been going on ever since, but it is a racial Becoming, technically called *Phylogeny*, very different from the individual Becoming, technically called *Ontogeny*. Yet we must not think of development and evolution as unrelated. For, in the first place, every animal tends to climb up its own genealogical tree, the individual development being to some extent a much shortened recapitulation of the great steps that occurred in the racial evolution. This we see in the young tadpole when it has for a short time the type of heart and circulation that is characteristic of fishes. And, in the second place, at the beginning of each individual development there may be shufflings of the hereditary cards that result in novelties or variations that may form part of the raw material of future evolution. But our present point is simply to make it clear that two good words should not be mixed up with one another: the *development* of a fertilised egg into a hen is a well-known individual Becoming, very different from the dimly known *evolution* of the class of Birds from a reptilian ancestry, onwards—a racial Becoming.

This brings us now to *Organic Evolution* in particular, which has been going on for hundreds of millions of years and has led to the plants and animals that are living around us to-day. Perhaps it is better to say explicitly that the organic evolution of past ages has led on not only to our present-day fauna and flora, but also to their relations with one another and with their inanimate environment. For besides the evolution of the diverse forms of life, there has been an evolution of their linkages with one another and with their surroundings. In short, there has been an evolution of what we called in the Eighteenth study " the web of life."

Organic Evolution may be defined as a natural process of racial change in a definite direction, in the course of which new individuals emerge, and are established alongside of or in place of the stock from which they arose. When several different parts of the plant or animal are changing or have changed at once, it will be necessary to say " in a definite direction, or in several definite directions."

This definition, which we here suggest, should be read over several times, for it cannot but be full of difficulties. Let us add two or three explanations. The first plants and animals were single cells, and when we compare single cells with, say, the cedars of Lebanon or with the eagles soaring above them, we cannot help using some word like *progress*. There has been an advance of life, marked by increasing complexity (differentiation) and by increasing control (integration) ; and in the animal world there has been an increasing emancipation of mind. Yet evolution has not always been progressive, thus parasitic animals and many sedentary animals have slipped backwards. But even those organisms that have retrograded have become better fitted to the immediate conditions of their life ; thus the tapeworm in the golden eagle's food-canal is " fit " for its inglorious lot. On the whole, however, organic evolution has been progressive ; and we cannot forget that it has led to ourselves !

The French word transformism was often used in Darwin's day to express the idea of Organic Evolution, but it is misleading in so far as it suggests that one type changes into another, say a reptile into a bird. There are certainly many surprising changes in individual development, as when a caterpillar becomes a butterfly, or a tadpole a frog, but that is not the way in which new types arise or arose in evolution. In a particular kind of species of plant or animal offspring appear that are a little different from their parents ; they are new departures, variants or mutants, and their peculiarity may give them a good chance of surviving, perhaps a better chance than their parents had. This may be particularly noticeable if they move or are carried into slightly different surroundings, or if the old surroundings change. If the novel peculiarities are handed on to the next generation, and are perhaps increased in amount ; and if this goes on for a long time, the variants may come to form a true-breeding new species. But their success does not necessarily involve the wiping out of the original stock, for the old and the new may flourish side by side. Yet if circumstances changed it may quite readily come about that the old species goes to the wall, and the new one survives. But this is a very difficult picture from that suggested by the word transformism, which leads us to think of one type itself changing into another. The wild Rock Dove, the ancestor of all our domestic pigeons, did not itself change into a Fantail, but the stock of Rock Doves gave rise by variation or mutation to new departures that were the first steps in the series of changes that led to the firm establishment of a true-breeding race of Fantails. It must be noted, however, that the offspring sometimes differ brusquely and considerably from their parents. There are big mutations as well as small fluctuations; to speak metaphorically, the Proteus of life leaps as well as creeps.

The general idea of Organic Evolution is a way of looking at the origin of the different kinds of living creatures. It suggests that the familiar plants and animals around us arose in the course of generations of slow change from ancestors a little different, and so on—back and back—to very simple aboriginal ancestors. It suggests that there have been in Nature processes of gradual change similar to those by which the races of domesticated pigeon arose from the wild Rock Dove, or the various races of wheat—so precious to mankind—from a wild mountain grass, such as still grows on the shoulders of Mount Hermon. This evolutionist way of looking at the origin of present-day plants and animals is the only scientific outlook, but, from the nature of the case, it cannot be logically demonstrated as we might demonstrate the Law of Gravitation or the Law of the Conservation of Energy. Yet it is possible to illustrate the kind of evidence by which Darwin convinced all thoughtful men that the evolutionist view was reasonable. We shall take birds as our illustration.

What evidence is there in support of the conclusion that birds, the conquerors of the air, evolved from an extinct stock of reptiles? It is still uncertain whether their origin is to be looked for in an order of Dinosaurs called Ornithischia or in another order called Pseudosuchia, but that is a question for experts. Let us keep to the big conclusion accepted with practical unanimity by zoologists, that birds emerged from extinct bipedal reptiles. What evidence is there of this ?

First of all, there is what might be called historical evidence. Two specimens of the oldest known fossil bird, Archæopteryx by name, are so beautifully preserved on the fine-grained lithographic stones of Bavaria that every part of the skeleton is quite clear except the breastbone. This very interesting " first bird," about the size of a crow, was in many ways reptile-like. Thus it had teeth in both jaws, a long tail like a lizard's, and a half-made wing with three clawed digits. In short, it is a very good example of a connecting-link.

In the second place, every bird carries about in or on its body indelible marks of its reptilian ancestry. The pieces of a compound bill, well seen in the albatross, correspond to the scales on a reptile's jaws, and the scales on the bird's toes also betray its ancestry. The annual shedding of the outer parts of a puffin's quaint bill recalls the periodic shedding of the outermost layer of the epidermis in reptiles, and so does the autumnal moulting of the claws of the toes in a grouse. The lower jaw of a mammal is one bone on each side, and works on an upper bone called the squamosal, but birds and reptiles agree in having a lower jaw made up of five or six bones on each side, and articulating with an upper bone called the quadrate. In fact, the body of the bird is swarming with reptilian features.

In the third place, there is evidence from development. The egg of a goose is very like the egg of a crocodile, both inside and out ; and for some days the development of the embryo bird is much the same as that of the embryo reptile. For a time the developing bird and the developing reptile are travelling, as it were, together along the same highway, but towards the end of the first week of brooding the two embryos, have, so to speak, parted company, each pursuing its own path. Even then, however, in some parts of its body, such as the ankle-joint, the heart, and the brain, the developing bird recapitulates certain steps taken by the developing reptile.

We need not give other examples, for the fact is that all the features of all plants and animals become evidences of evolution when we know enough about them. The evolution-idea is a master-key that fits every lock in which it can be inserted; and without the evolution-idea the world is scientifically unintelligible.

But another and more difficult question now rises: What have been the factors that have operated in the long-drawn-out process of organic evolution, and still continue to operate. Our knowledge of these factors is still very young, and the problems are undoubtedly very difficult. We know that birds must have emerged from an extinct reptilian stock, but we are very uncertain in regard to the factors by which the advance was brought about. In a general way, however, it may be said that organic evolution is a process of *varying* and *entailing*, *sifting* and *singling*. Let us think of these factors for a little.

Variation is the origin of something new that makes offspring a little different from their parents and from one another. In some way or other there are changes in and new arrangements of the hereditary qualities at the very beginning of life, in the egg-cell and the sperm-cell, and when they come together in fertilisation. There may be a little more of one quality, say swiftness, and a little less of another quality, say heaviness of skeleton. Or there may be a more marked novelty, a new pattern or originality, such as horns, crests, spurs, wattles, colour-change, death-feigning, "homing" power, and sheer cleverness. Many of the novelties are negative in character, such as absence of pigment in albinos, as in white blackbirds, or the absence of a tail, as sometimes happens in kittens. These inborn changes are called variations, and they include graduated *fluctuations* (a little more or a little less) and brusque *mutations*, such as the sudden appearance of a Greater Celandine with its leaves cut up into fine strips. Mutations in particular are very likely to be continued in the next generation. When they come, they come to stay; and they illustrate what is called Mendelian inheritance, as explained in our Nineteenth study. It is important to try to distinguish "variations" in the strict sense from "modifications" which are impressed on the body as the direct result of some peculiarity in surroundings, food, or habits. These modifications are imprints not outcomes, and may be illustrated by the tanning of a white man's skin in the course of years of work in the Tropics, or the change of colour in a canary's plumage when it is fed on peculiar food, or the strengthening of muscle when it is much exercised. There is very little evidence to show that these modifications can be to any extent handed on to the next generation.

If variability is the first great factor in Organic Evolution, heredity is the second. As we have explained in our Nineteenth study, heredity is the flesh-and-blood linkage between successive generations, and it secures the handing on of the characteristic features that mark one kind of living creature from another. It is like the First Law of Motion: it means continuance. It secures that the past shall live on in the present, and in the future too. It may be compared to what is called entailing a landed estate, a legal arrangement by which a father secures that his property will pass on to his heir, and to his heir's heir.

A third factor in organic evolution is selection or sifting, which includes all the different ways in which living creatures are discriminately winnowed. It operates whenever an organism is appreciably more successful than its fellows in virtue of some quality that it has and they have not. It does not necessarily mean that those without the advantageous new feature are immediately killed off, though that sometimes happens. It includes gentle as well as severe sifting, and it will reach the same end—*the survival of the relatively more fit*—by allowing certain variants with a profitable feature a longer life and a larger or more successfully launched family, or by stringently cutting off those that are without the feature in question.

A fourth factor in Organic Evolution is isolation, a technical term for all the ways of narrowing the range of intercrossing among variants of the same species. Thus on each of six islands of the Galapagos group there is a particular species of giant tortoise, nearly related to one another and doubtless all descended from one. The separate islands are the volcanic peaks of a submerged peninsula, and as the tortoises do not swim, variants that appeared on the different islands have bred with others like themselves and six or more stable species have resulted. Similarly there is a peculiar Orkney vole and a peculiar St. Kilda wren. Isolation brings about in-breeding, and this tends to fix characters.

So far we have spoken of Organic Evolution in general, but we think that it makes for clearness to keep *human* evolution by itself. For man has reason, whereas animals are not known to rise above intelligence; man alone has language in the strict sense; man is more or less aware of his past history, and in the light of ideals he can to some extent control his future. Higher animals sometimes work towards a concrete personal end, something they probably picture, something with a strong desire, but they never "make history" as men sometimes do. Furthermore, man has learned to register many of his gains outside himself in traditions, institutions, literature, art, and so on, in a way that is only hinted at in the ant-hill and the beaver-village.

Perhaps we should go a step farther and distinguish *social* evolution, with particular reference to those cases where the human activity rises beyond the individual and involves the corporate endeavour of a community or a society. There can be no doubt that the progress of mankind depends in great part on an increase in man's practical sensitiveness to the advances that are registered and continued outside himself altogether, in what may be called the "social heritage." But modern science has more unsolved problems in regard to human society than in regard to any other order of facts. So, in an introductory book like this, it is perhaps wisest to say no more at present.

INDEX

www.ingramcontent.com/pod-product-compliance
Lightning Source LLC
LaVergne TN
LVHW011126300926
849160LV00008B/535
* 9 7 8 1 1 3 8 5 6 7 7 4 0 *